T0289989

Climate Change: Today and Tomorrow

Climate Change: Today and Tomorrow

Edited by Dustin Doyle

SYRAWOOD
PUBLISHING HOUSE

New York

Published by Syrawood Publishing House,
750 Third Avenue, 9th Floor,
New York, NY 10017, USA
www.syrawoodpublishinghouse.com

Climate Change: Today and Tomorrow
Edited by Dustin Doyle

International Standard Book Number: 978-1-64740-344-7 (Hardback)

Cataloging-in-publication Data

Climate change : today and tomorrow / edited by Dustin Doyle.
    p. cm.
Includes bibliographical references and index.
ISBN 978-1-64740-344-7
1. Climatic changes. 2. Climate change mitigation. 3. Climatology. I. Doyle, Dustin.
QC903 .C55 2023
577.22--dc23

Table of Contents

Permissions

List of Contributors

Index

PREFACE

Climate refers to a long-term pattern of weather in a particular area. Temperature, humidity, atmospheric pressure, precipitation, and wind are the common meteorolgical variables that are measured to classify climatic conditions. The climate system has five components which include atmosphere, hydrosphere, cryosphere, lithosphere and biosphere. A location's climate is influenced by its altitude, latitude and geography along with nearby water bodies and their currents. Climate change is characterized by changes in the global and regional climates over time. It reflects variations in the atmosphere's variability or average condition over different time spans ranging from decades to millions of years. The climate is dynamic in nature and it keeps changing. Along with human activity, natural processes such as the Earth's tilt, volcanoes, comets and meteorites, continental rifts, and ocean currents can cause these changes. This book unravels the recent studies on climate change. It will serve as a valuable source of reference for graduate and postgraduate students.

This book has been the outcome of endless efforts put in by authors and researchers on various issues and topics within the field. The book is a comprehensive collection of significant researches that are addressed in a variety of chapters. It will surely enhance the knowledge of the field among readers across the globe.

It gives us an immense pleasure to thank our researchers and authors for their efforts to submit their piece of writing before the deadlines. Finally in the end, I would like to thank my family and colleagues who have been a great source of inspiration and support.

Editor

1

Retooling Smallholder Farming Systems for Climate Change Resilience Across Botswana Arid Zones

Nnyaladzi Batisani, Flora Pule-Meulenberg, Utlwang Batlang, Federica Matteoli and Nelson Tselaesele

Contents

N. Batisani (✉)
Botswana Institute for Technology Research and Innovation, Gaborone, Botswana

Food and Agriculture Organization of the United Nations, Rome, Italy

F. Pule-Meulenberg · U. Batlang · N. Tselaesele
Botswana University of Agriculture and Natural Resources, Gaborone, Botswana
e-mail: fpmeulenberg@buan.ac.bw; ubatlang@buan.ac.bw

F. Matteoli
Food and Agriculture Organization of the United Nations, Rome, Italy
e-mail: Federica.matteoli@fao.org

Abstract

Background: Scientific progress and developments in technology have improved our understanding of climate change and its potential impacts on smallholder farming systems in sub-Saharan Africa (SSA). The persistence of such smallholder farming systems, despite multiple exposures to climate hazards, demonstrates a capacity to respond or adapt. However, the scale and intensity of climate change impacts on smallholder farming systems in SSA will overwhelm any indigenous coping mechanisms developed over centuries. Therefore, there is need to co-develop resilient farming systems with farmers and extension workers in anticipation of the looming food security challenges in the midst of climate change.

A survey comprising of participatory rural appraisal, focus group discussions, participatory resource mapping, and SWOT analysis was carried out for the purposes of farming systems diagnosis in reference to their resilience to climate change in three districts cutting across dry arid zones of Botswana agricultural landscape. The survey also sought to identify vulnerability of the farming systems to climate change and subsequently co-develop with farmers and extension workers new climate proofed farming systems.

Results: Detailed evaluation of current systems and their strengths and weaknesses were identified. Farmers highlighted constraints to their production being mainly drought related but also lack of production inputs. These constraints are location and context specific as extension areas within a district highlighted different challenges and even different CSA practices for similar production constraints. Through participatory approaches, farmers were able to identify and rank potential climate-smart agriculture practices that could ameliorate their production challenges and subsequently developed implementation plans for these practices.

Conclusions: The study demonstrates that climate change is already having significant adverse impacts on smallholder farming systems and therefore, climate proofing these systems is necessary if livelihoods of smallholder farmers are to be sustained. Therefore, retrofitting current farming systems to be climate resilient is the first step to climate proofing smallholder farmers' livelihoods.

Keywords

Botswana · Farming systems · Adaptation strategies · Climate change · Smallholder farmers

Introduction

Agriculture is a proven path to prosperity as no region of the world has developed a diverse, modern economy without first establishing a successful foundation in agriculture. This trend is going to be critically true for Africa where, today, close to 70% of the population is involved in agriculture as smallholder farmers working on parcels of land that are, on average, less than 2 hectares. As such, agriculture remains Africa's route for growing inclusive economies and creating decent jobs mainly for the youth. In Botswana, the agriculture economic sector is limited mainly to range resource-based livestock and pockets of arable farming based on rainfall and limited irrigated agriculture at several places (Alemaw et al. 2006). However, despite this aforementioned and the recent slowing in economic growth across much of the continent mainly due to the sharp drop in the global prices of oil and minerals, the prospects for African agriculture looks favorable. The African food market continues to grow with World Bank estimates showing that it will be worth US$1 trillion by 2030 up from the current US$300 billion (World Bank 2007, 2010). Demand for food is also projected to at least double by 2050. These trends, combined with the continent's food import bill, estimated at a staggering US$30–50 billion, indicate that an opportunity exists for smallholder farmers – Africa's largest entrepreneurs by numbers – who already produce 80% of the food to finally transition their enterprises into thriving businesses (AGRA 2017). Botswana is a net food-importing developing country (NFIDC); thus there is an opportunity to increase domestic production of basic foodstuffs, particularly cereals (grain sorghum and maize) and pulses. The national demand for cereal stands at 200,000 t per year, of which only 17% is supplied through local production (GoB 2019). Therefore, investments in arable agriculture will stimulate private sector development, create employment, value-addition opportunities, and enhance food security and ultimately exports.

Climate change poses a challenge to the attainment of agricultural potential. Climate change is threatening to undo decades of agricultural development efforts in developing countries with scientific projections pointing to a warmer climate characterized by increases in both intensity and frequency of extreme climate events, particularly in sub-Saharan Africa. Nkemelang et al. (2018) observed an increase in extreme weather events such as heat waves and late and high spatial and temporal rainfall variability across Botswana. Therefore, although climate adaptation is a global requirement, the need for adaptation is considered higher among developing countries where vulnerability is presumably higher (Adger 2003) and also in the interest of individual farmers who rely on the revenue generated from agricultural production (Holzkämper 2017). The need for adaptation is especially true in Africa as the population is highly dependent on rainfed agriculture (the most climate-sensitive sector) and particularly for smallholder farmers as they generally have limited adaptive capacity (Morton 2007), hence considered among those who will suffer most from the impacts of climate change (Easterling et al. 2007). Mogomotsi et al. (2020) highlight the vulnerability of smallholder rainfed famers to climate change and variability in Botswana. Smallholder agriculture has long been characterized by adaptive and flexible strategies to reduce vulnerability to climate natural variability and soil depletion (Adger et al. 2003; Tschakert 2007; Thomas et al. 2007; Eriksen et al. 2008). African farmers, particularly in dry land areas, have developed both on- and off-farm adaptation strategies in response to reoccurring droughts. Cooper et al. (2008) observed that coping better with current climatic variability in rainfed farming systems of sub-Saharan Africa is an essential first step in adapting to future climate change, while Muyambo et al. (2017) highlight the role of indigenous knowledge in drought risk reduction.

Nevertheless, climate change is going to negatively affect smallholder farmers beyond their coping capacity for naturally-occurring droughts, hence the need to hybridize traditional drought coping mechanisms with technology through co-development of climate resilience farming systems involving farmers, extension workers, and agricultural experts. Hence the need to retool farmers and extension officers for climate. Williams et al. (2019) observed that smallholder systems heterogeneity requires local specific climate adaptation for reducing the negative impacts of changing climate in regions heavily relying on small farms agriculture.

Due to the instability of agricultural production as a result of complex, dynamic, and interrelated factors such as climate, markets, and public policy that are beyond farmers' control, there is a need for farmers to develop new farming systems that incorporate innovations in their objectives, organization, and practices adapted to changing production contexts (Martin et al. 2013). While Thornton et al. (2018) highlighted that the scale of change required to meet the sustainable development goals, including those of no poverty, zero hunger, and the urgent action needed to address climate change, will necessitate the transformation of local and global food systems. Thus the main aim of this chapter is to develop resilient farming systems in Gantsi District and Bobirwa and Boteti sub-districts of Botswana. The chapter is premised on the following specific objectives: to evaluate current farming systems in the three districts and identify climate-smart practices within each district; identify alternative livelihood options (off-farm) in the catchment areas; identify indigenous knowledge of agricultural practices (ethno-veterinary and ethno-botanical) to cope with effects of climate change and co-identify (farmers and extension workers); and recommend potential climate-smart agricultural practices. The Republic of Botswana is a landlocked country with an area of 582,000 km^2 (Fig. 1). The climate is semi-arid to arid with high spatial rainfall variability. Rainfall decreases and temperature range increases westward and southward, varying from 650 mm per annum in the east to 230 mm in the south-west.

Agriculture plays a significant role in the lives of rural communities where it provides food and income and employs a majority of the rural inhabitants. However, the Intergovernmental Panel on Climate Change (IPCC 2007) has identified the country as vulnerable to climate change and variability, probably due to its low adaptive capacity and sensitivity to many of the projected changes. Therefore, it implies that with climate change, including high temperatures and frequent droughts conditions for agriculture, will worsen. For example, models have predicted that parts of Botswana will become much drier and hotter (IPCC 2001). Currently, there are projected statistically significant decreases in mean rainfall and increases in dry-spell length at each global temperature level (Nkemelang et al. 2018). IPCC special report on global warming of 1.5°C underlined that areas in the south-western region, especially in South Africa and parts of Namibia and Botswana, are expected to experience the largest increases in temperature (Engelbrecht et al. 2015; Maúre et al. 2018).

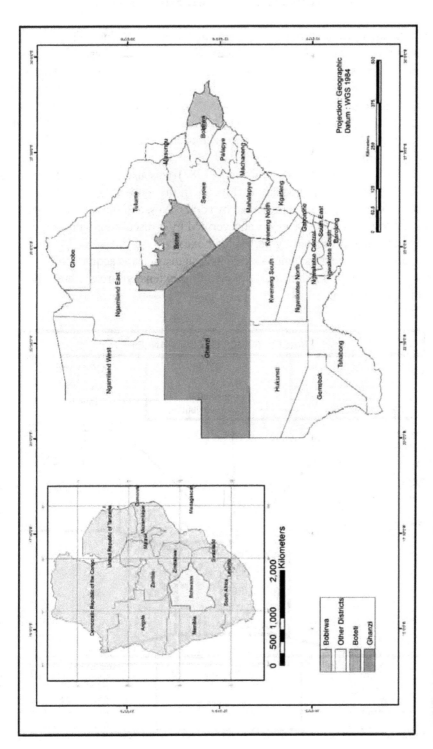

Fig. 1 Botswana location and the three study districts

Gantsi District and Bobirwa and Boteti Sub-Districts

Although the three areas of interest are very similar in their vulnerability to climate change impact, they vary in terms of land area, population, and physiography. Gantsi District has mostly sandy, infertile soils with low water holding capacity. Farming is the dominant economic activity in the district with pastoral being most dominant although of recent there has been an increase in irrigated agriculture. Bobirwa sub-district has relatively fertile soils although some. Integrated pastoral and arable rainfed farming are major activities in the district. Boteti sub-district practices both arable and pastoral farming.

A farming system model by Collinson (1987) was used to aid with farming systems diagnosis of the three districts (Fig. 2). In the model, there are visible and observable aspects, which are represented by solid boxes, and those that are deduced from the description of these aspects and verified by a discussion with farmers and extension officers, represented by perforated boxes.

Information on current farming systems in each district was acquired through key informants, farmers, and extension workers. Two extension areas were randomly selected in each district.

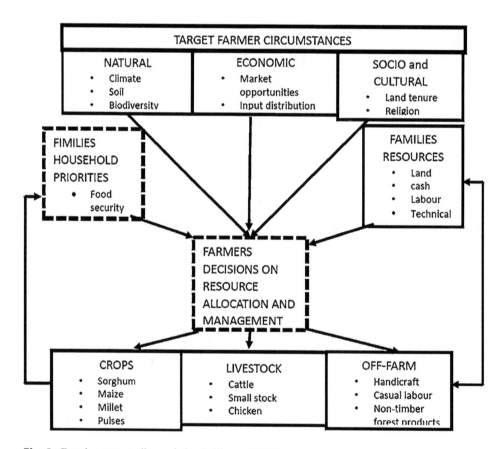

Fig. 2 Farming system diagnosis by Collinson (1987)

(i)   Gantsi District (Charles Hill, Ncojane)
(ii) Boteti sub-district (Mosu, Moremaoto)
(iii) Bobirwa sub-district (Bobonong, Gobojango)

Multistage sampling technique was used to determine the ideal sample size of participants in the focus group discussion (FGD) using Krecjie and Morgan (1970) formula. The multistage sampling process was conducted with the assistance of extension areas workers who provided facilitators with the number of active farmers in their catchment. Triangulation was applied where specific information was needed. A SWOT analysis was used to evaluate current farming systems, their strengths (resilient to climate change and droughts), and weakness (vulnerability to climate change and droughts).

The sharing of ideas (experience) among farmers and extension workers allowed them to evaluate the shared experience and seek more solutions to the prevailing livestock and crop production challenges and CSA practices in their respective extension areas. The list of proposed CSA options was discussed with participating farmers to ensure a common understanding was reached between them, facilitators and extension workers. During the discussion, farmers were encouraged to come up with additional practices especially those relating to indigenous knowledge practices (ethno-botanical and veterinary practices). At the end, ranking of the proposed CSA practices was done according to Khatri-Chhetri et al. (2017).

The discussion of the predetermined options was designed into a participatory action plan that addressed the resources needed for implementation of the ranked CSA potential practices. The action plan and the developed activity plan was a step-by-step process that helped these groups of farmers together with their extension workers and facilitators to design and deliver solutions to address proposed climate-smart agricultural practices. Farmers ranked potential CSA practices for developing climate-resilient farming systems and together with extension workers developed their implementation guidelines. Figure 3 is a summary of the methods that were used to collect data needed to co-develop climate-smart resilient farming systems across the study areas.

Facilitators used a stated preference method to analyze the farmers' preference of CSA practices. In the stated preference method, participating farmers were asked about their preferences in a list of practices (Khatri-Chhetri et al. 2017).

Current Farming Systems in Gantsi District and Bobirwa and Boteti Sub-Districts of Botswana

Table 1 displays CSA practices currently found in the three districts. Such practices can be divided into crop and livestock related. Crop technologies that were common to all areas included fodder production and intercropping, whereas disease control and public health as well as ethno-veterinary practices were the only two livestock practices that cut across the three districts. Fodder production included planting of cowpea, lablab, maize, melons, forage sorghum, napier grass, and soybean. Cowpea

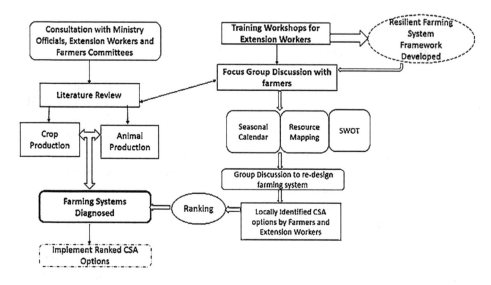

Fig. 3 Flow of methods used in developing the farming systems

and maize were not specifically planted as fodder, but their stover is normally fed to livestock after harvesting. Intercropping involves planting of rows of cereals such as maize, sorghum, and millet alternating with legumes such as cowpeas, lablab, and soybean and is part of integrated pest management as it provides conducive environment for natural enemies (Obopile et al. 2018) and also in maintaining soil fertility through biological nitrogen fixation (Pule-Meulenberg et al. 2018).

Current farming systems in Gantsi District is predominantly livestock farming as observed from the large number of cattle posts (livestock holdings) (Fig. 4). The rearing system is free ranging based on rainfed natural pastures. This resource (natural pastures) is prone to droughts and overgrazing leading to a large number of livestock mortalities.

As an adaptation move, farmers are moving toward some integrated crop-livestock systems with production of fodder crops, while some large commercial cattle ranchers use parts of their farms for diversifying into vegetable production. Smallholder farmers are also practicing small-scale irrigated gardens. Maize and cowpeas are the main rainfed arable crops in the area, and because of low drought tolerant of maize, yields are low in most years (Fig. 5). Minimum tillage is practiced and pioneered as a conservation agriculture (CA) in the district. However, due to the sandy nature of the soil, farmers complained of the disappearance of planting pits/ basins, making their construction a laborious task that has to be repeated every year.

Boteti sub-district just like Gantsi District is a predominantly cattle-rearing area although has more rainfed arable farming. Figure 6 shows the spatial distribution of land uses in the sub-district. The main crops grown are maize, cowpeas, sorghum, and millet; the latter two being drought-tolerant crops, whose limitation is susceptibility to bird damage. Of recent, crop damage by wildlife especially elephants have increased in the sub-district.

Table 1 CSA practices currently found in the Gantsi District and Bobirwa and Boteti sub-districts

Current CSA practices	CSA details	Gantsi District	Bobirwa Sub District	Boteti Sub District
Planting in tires	Vegetables	x		
Irrigation	Combination of different types of vegetables	x		
Holistic livestock management	Planned grazing in fenced farm	x		
Fodder production	Cowpea, lablab, Maize, Melons, Forage sorghum, sugar cane, soya	x	x	x
Intercropping	Intercropping legumes and cereals	x	x	x
Crop rotation	Cereals rotated with legumes	x		x
Backyard gardening	Different types of vegetables	x		
Integrated farming	Combination of different crop enterprises	x		
Integrated farming	Combination of different types of livestock, fodder, vegetables, field crops, and poultry	x		
Supplemental feeding	Using commercial and fodder	x	x	
Disease control and public health	Construction pit latrines to curb the spread of beef measles	x	x	x
Poultry production	Constructed poultry houses which control temperatures using natural ventilation. Collection of chicken manure to sell for vegetation production		x	
Small stock production	Apply recommended management practices that follow small stock calendar		x	x
Production of marula oil cake	Marula by products used for animal feeds		x	
Utilization of indigenous tree species for livestock feeding	Cutting *Vachellia* spp. (*Vachellia tortilis* and *V. erioloba*) species and mixing with stover to make livestock feeds		x	
Pigs production	About 300 individual pigs housed in large paddocks		x	
Minimum tillage (ripping using tractor)	Mainly cowpeas, watermelons, maize, sorghum			x
Minimum tillage (ripping using donkeys)	Cowpeas, watermelons, maize, sorghum			x
Planting pits/basins	Cowpeas, watermelons, maize melons, maize	x		x
Cover cropping	Mainly cowpeas, pumpkins, lablab, watermelons, melons		x	x
Fertilizer application as per soil analysis	Use of fertilizers on sorghum, maize			x

(continued)

Table 1 (continued)

Current CSA practices	CSA details	Gantsi District	Bobirwa Sub District	Boteti Sub District
Kraal manure application	Application of kraal manure on maize			x
Provision of housing	Housing for goats' kids			x
Vaccinations	Vaccinate against livestock and poultry diseases	x	x	x
Use of solar energy	Solar panels to generate electricity			x
Ethno veterinary practices	Examples: wood ash for retained placenta; powdered dead/sun bleached millipede skeleton to treat eye infection	x	x	x
Use of tolerant breeds	Use of Tswana goats, Boer goats			x

Figure 7 displays current farming systems for crops and livestock in Letlhakane extension area of Boteti sub-district. In order to minimize drought impacts on both livestock and crops, it is imperative to increase the number of people that practice climate-smart agriculture technologies such as conservation agriculture, fodder production, and the use of drought-tolerant germplasm.

Of the three districts, in terms of both physiography and farming systems, Boteti lies at the transitional zone between Gantsi and Bobirwa in having some aspects of sandy predominant in Gantsi but also having the more fine-textured fertile soils found in Bobirwa. Similar to Gantsi District, livestock production is dominant in the Boteti sub-district. In Boteti sub-district, rainfed arable agriculture comprising sorghum, millet, maize, and cowpeas are among the major crops grown similar to Bobirwa.

Alternative Livelihoods in Gantsi District and Bobirwa and Boteti Sub-Districts

Table 2 shows alternative livelihoods outside the agricultural sector. It is evident that Gantsi District had many more alternative livelihood activities compared to the two sub-districts of Bobirwa and Boteti. This is probably due to the fact that Gantsi township is more urban compared to the villages of Bobonong and Letlhakane, which are the sub-district headquarters for Bobirwa and Boteti sub-district, respectively. In each of the three study areas, livelihood activities were varied as shown in Table 2. The common activities that cut across the three districts included employment through the Ipelegeng program (a Government of Botswana poverty alleviation scheme), sale of veld product, running of tuckshops, domestic work (being a maid), taxi services, sale of traditional brews, and catering services.

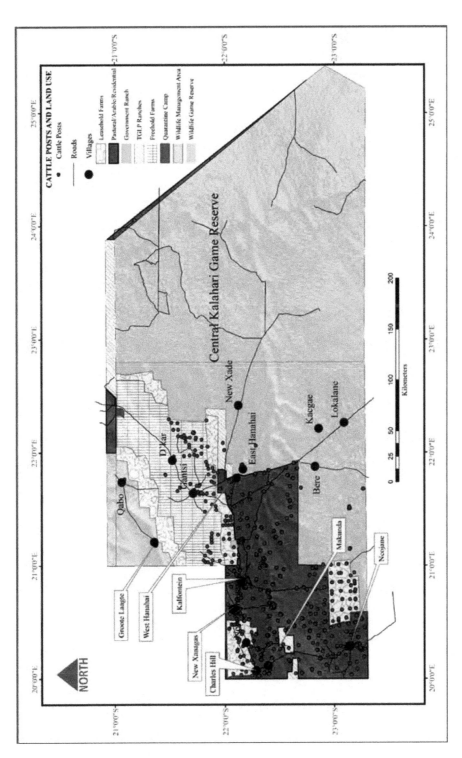

Fig. 4 Spatial distribution of current farming systems in Gantsi District

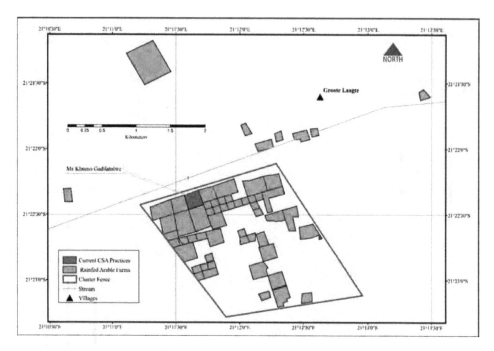

Fig. 5 Current farming systems for crops and livestock in Groote Laagte extension area (Gantsi)

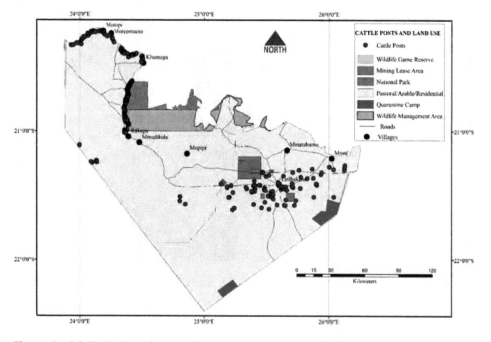

Fig. 6 Spatial distribution of current farming systems in Boteti district

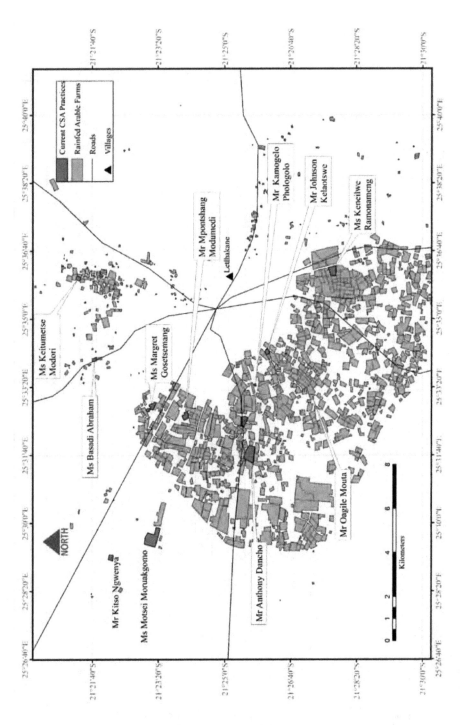

Fig. 7 Current farming systems for crops and livestock in Letlhakane extension area (Boteti)

Table 2 Alternative livelihoods for Gantsi District and Bobirwa and Boteti Sub-Districts

Alternative livelihoods	Gantsi District	Bobirwa sub-district	Boteti sub-district
Employed in Ipelegeng projects	x	x	x
Traditional herbalists selling traditional medicines	x	x	
Selling of veld products	x	x	x
Provision of mechanical services mainly in Gantsi township	x		
Renting out houses	x		
Performing arts	x		x
Sale of traditional brews in shebeens	x		
Micro-enterprises (tuckshop)	x	x	x
Sales agents	x		
Farm work (livestock herding)	x		
Horse racing (jockeys)	x		
Dog racing	x		
Domestic work	x	x	x
Mobile safari	x		
Sale of game meat from community trusts	x		
Taxi services	x	x	x
Hired donkey carts for transport	x		
Game farming	x		
Handicrafts	x	x	x
Traditional brews	x	x	x
Labor for destumping arable fields	x		
Illicit substances (marijuana)	x		
Provision of transport for goods	x		
Cattle sales agent	x		
Brick molding	x		x
Poaching and hunting	x		

Activity			
Bakery	x		
Dependency on male/female partners	x		
Garbage collection	x		
Fishing		x	x
Harvesting wood, grass and poles		x	x
Catering services	x	x	x
Sewing (tailoring)		x	x
Pottery			x
Thatching		x	x
Food processing			x
Sale of secondhand clothes		x	x
Sale of ornamental and fruit trees		x	
Sale of semi-precious stones		x	
Using recycled paper to make flower pots		x	
Sale of empty cans		x	
Recycling old tires to make livestock troughs		x	
Leatherworks		x	
Hair salons			
Wood carving		x	x
Destumping to increase hectarage in fields		x	
Panel beating and motor mechanic		x	
Pottery			x
Welding			x
Upholstering			x

The types of veld products sold are area specific although there are some common ones. For example, they include the morama bean (*Tylosema esculentum*), mongoose seed (*Bauhinia petersiana*), and wild berries such as *Grewia flava*. In Bobirwa, some of the veld products being sold are the mopane worm (*Imbrasia belina*), fruit of baobab (*Adansonia digitata*) tree, marula (*Sclerocarya birrea*) fruits, monkey orange, *Grewia flava* berries, *Vangueria infausta* fruits, *Mimusops zeyheri*, and wild vegetables such as *Cleome gynandra*, *Amaranthus* spp., and wild okra. For the Boteti sub-district, veld products include the harvesting of the water lily which is as a condiment for meat and sale of indigenous vegetables as described for Bobirwa and wild fruits as baobab (*Adansonia digitata*) and marula (*Sclerocarya birrea*). The sale of firewood and handicrafts was also common among the study areas, the difference being the type of species that is being sold.

It is noteworthy that veld products are dependent on climate because during prolonged droughts when crops fail and livestock die, wild plants are equally affected; hence such products go off the markets. This situation is also true for traditional brews that are either made from mainly sorghum or wild fruits. During drought years, they would also not be available. This scenario shows the extent to which smallholder farmers are exposed to the effects of climate change, hence the need to adopt climate-smart technologies, on and off farm.

The recurring droughts leading to asset losses have driven communities in these three districts to develop some alternative livelihoods strategies for coping with drought challenges. Nevertheless, these alternative livelihoods strategies developed to cope with natural droughts are unlikely to match the severity and frequency of climate change-induced droughts because of their short-term nature (Table 2), hence the need for coming up with CSA technologies by Government and other stake-holders. However, the adoption of these CSA technologies has not been promising due to among other obstacles, weaknesses as highlighted in Table 3. These weak-nesses or malalignment of the current CSA technologies used in the districts or their implementation may be due to that they were never meant for farming systems and cultures found in the three districts, therefore could have been adapted for these systems before implementation. Mwongera et al. (2017) noted that approaches that aim to identify and prioritize locally appropriate climate-smart agriculture technolo-gies will need to address the context-specific multidimensional complexity in agricul-tural systems.

Therefore, to improve the adoption of CSA technologies, there is need for co-development of technologies with farmers and extension workers at the local level. Through participatory approaches, farmers ranked various CSA interventions. The examples in Table 4 below shows that preferences vary at local extension levels within a district and also between districts, highlighting the importance of consider-ing local context and dynamics of farming systems when designing CSA practices and interventions (Tables 4–9). Khatri-Chhetri et al. (2017) noted that farmers' priorities for CSA technologies are linked with prevailing climatic condition of particular location, socio-economic characteristics of farmers, and their willingness to pay for available technologies.

Table 3 SWOT analysis of current CSA farming practices in Gantsi District and Bobirwa and Boteti sub-districts

Strengths	Weaknesses
Improved soil fertility	Use of hybrid seed destroyed traditional crop varieties
Technologies are adaptable to climate change	Destruction of CSA crop structures by wind erosion and livestock
High carbon sequestration	Unavailability of equipment for CSA crop practices
Improved food and nutrition security	The decision to sell livestock is predominantly done by males
Both males and females participate in farming activities	The decision to sell crops is predominantly done by females
Availability of underground water	Shortage of livestock grazing land and grazing resources
Availability of arable land	Poor vegetation
Supportive Government policies	Shortage of farming equipment
Mild winters	Lack of farm input suppliers
Availability of veldt products (natural resources)	Poor roads and infrastructure
Fertile soils	Lack of research and development (e.g., abattoirs)
Moisture conservation	Long-time taken by Land Board to allocate land
Pest and weeds control (break life cycle)	Shortage of labor
Reduction of soil erosion	Very interested to do fish farming
	No financial assistance for fisheries
	Few trained officers in Aquaculture
	Unreliable rainfall
	Reliability on Government hand out
	Conflict on Government policies
	Unwillingness to change by farmers to new technologies
	High farmers to extension worker ratio
	Lack of documentation on fish farming in Botswana
	Blanket application of fertilizers

Opportunities	Threats
Adaptation with co-benefits	Occurrence of extreme weather conditions (e.g., change of start of rainy season)
Resilience to drought due to irrigation	Emergence of new pests and diseases
Diversified crop production	Introduction of alien grass species in the rangelands (e.g., *Cenchrus biflorus*)
Employment creation	Competition of labor with Government programs
High demand for fish	Other farmers may not cooperate in rotational grazing and correct stocking rate in communal areas
Commodity clusters of farmers can be helping in farming systems (e.g., poultry and dairy)	Human-wildlife conflict (elephants)
Agro-tourism	To develop small stock slaughtering facility
Diversification in farming	Human-wildlife conflicts
Fish has a potential to grow since it is a new in Bobirwa sub-district and has a high nutritional value	The rain pattern has changed
	Heat waves
	Lack of raw materials for feeds
Enhanced import substitution	Use of hybrids seeds
Availability of sunlight for usage of solar power	Use of exotic livestock and poultry breeds
No Market for cattle to BMC because Bobirwa Sub District is not a green zone	Lack of buy in by the Government on the programs for fisheries (just like LIMID)
	High budget to start fish projects
Irrigation (crops, fodder production) almost the whole year	Invasive species
	Stock theft from the neighboring countries
	Aging farmers
Multiplication of Musi breed	Conversion of arable land to other land uses (e.g., tourism and settlements)
Improved soil fertility (legumes)	Contamination of environment

Table 4 Indigenous (ethno-veterinary) practices in the Gantsi, Bobirwa, and Boteti

Practices	Gantsi District	Bobirwa Sub District	Boteti Sub District
Use of charcoal for diarrhea in calves			x
Use of charcoal for dressing livestock wounds		x	
Use of charcoal to treat poisoning in dogs		x	
Use mix of dry donkey dung, salt, and wood ash and administered orally for retained placenta in small stock			x
Use of *Senna italica* (sebete) for the treatment of calf paratyphoid	x	x	x
Ximenia spp. for the treatment of foot rot in livestock			x
Use of sugar to treat visually impaired eyes and eye branding	x	x	x
Bandaging a fracture with wood and soft cloth		x	x
Use of *Aloe* spp. (mokgwapha) for foot rot			x
Use of wood ashes mix for snake bites			x
Use of cow dung after branding or dehorning			x
Thamnosma rhodesica (moralala) for prevention of miscarriage and still birth		x	x
Diospyros lycioides (letlhajwa) for control of *Pasteurella*			x
Use of *Aloe* spp. (mokgwapha) in birds against Newcastle disease and coccidiosis	x	x	x
Use of "thobega" against fractures in livestock		x	x
Dry old bleached millipede is crushed and applied to treat visually impaired eyes	x	x	x
Use of sugar to treat visually impaired eyes	x	x	x
Use of *Ziziphus mucronata* leaves to treat eye infections			x
Ziziphus mucronata pounded leaves and paste applied to dress wounds	x		
Use of wood ash mainly *Combretum imberbe* (Motswere tree) to control external parasites in chicken and puppies		x	x
Use of burnt cow dung against mosquitoes			x
Use of wild cucumber (mokapane) for the treatment of wounds			x
Use of dried cow dung smoke for the treatment of mastitis; burning cow dung is placed under the udder of the animal for the smoke to cover the udder			x
A string is tied on the warts and left until it falls off			x
Ash of *Vachellia mellifera* is mixed with water and administered orally when an animal has retained placenta; can also be used to control weevils in grain	x		x
Use of hot ring iron to treat eye infection			x
Cutting of tail and ears for treatment of animals against *Pasteurella*		x	x
Liquid paraffin external parasites on chicken, cats, dogs, and rabbits		x	

(continued)

Table 4 (continued)

Practices	Gantsi District	Bobirwa Sub District	Boteti Sub District
Potassium per manganite to control chicken diseases (e. g., coccidiosis)		x	
Use of brake fluid to control mites	x	x	
Use of bitter apple fruits (thontholwana/morolana) to treat eye infection		x	
Use of charcoal to treat poisoning in dogs		x	
Use of sugar for control of uterine prolapse	x		
Use of purslane to control uterine prolapse	x		

Indigenous Knowledge of Agricultural Practices in Gantsi District and Bobirwa and Boteti Sub-Districts

Farmers across the three districts indicated that they had indigenous agricultural practices pertaining to production of livestock and crops (Tables 4 and 5). Table 4 shows ethno-veterinary practices across the three districts. A number of practices such as the use of *Senna italica* to treat calf paratyphoid, treatment of eye infection in livestock using sugar and/or branding around the infected eye, treatment of poultry diseases such as Newcastle and coccidiosis using *Aloe* spp., and the use of dried up and bleached millipede carcass for eye infections in livestock were found in all the study areas. It is noteworthy that about a third of the practices are plant based. Consequently, prolonged droughts caused by climate change will affect them, hence the need to intentionally conserve wild plants. Other practices are based on plant products such as charcoal, wood ash, and smoke.

Table 5 displays the ethno-botanical practices found in the three districts. The use of a mixture of tobacco, chillis, garlic, onions, and soap or a combination of any two with liquid soap is common among the three districts for the control of aphids (*Aphis craccivora*), red spider mites (*Tetranychus urticae*), and fungi. The digging of trenches around arable fields for control against corn cricket (*Acanthoplus discoidalis*) and the traditional magical powers to protect fields against pest were common in the Gantsi District and Bobirwa sub-districts. The use of chilli blocks to scare away elephants where crushed chillis and mixed with cow dung, dried, and burnt was practiced in the Bobirwa and Boteti sub-districts. Much like with indigenous practices for livestock, about a third of the practices are plant based. Other practices such as digging of trenches to control corn cricket, animal snares, scare crows, use of reflectors to scare away elephants, and other can be referred to as physical.

It is interesting that out of the list of indigenous practices for both crops and livestock, none deal directly with increasing production but are all for the protection of pests and diseases. Furthermore, many of the practices are plant based and therefore are equally affected by the harsh effects of climate change.

Table 5 Indigenous (ethno-botanical) practices for crops in Gantsi District and Bobirwa and Boteti sub-districts

Practices	Gantsi District	Bobirwa Sub District	Boteti Sub District
Use of wood ash for pest control			x
Mixture of tobacco garlic and onion and sunlight for aphid control	x	x	x
Trenches for control of *Acanthoplus discoidalis* (setotojane)			x
Use of scare crows			x
Foot crushing of grasshoppers			
Animal traps (snares)			
Use of sugar to attract natural enemies			x
Application of *Combretum imberbe* and *Vachellia mellifera* (motswere and mongana wood ash, respectively, around plant stems and also application in seeds/grains to protect against storage pests			x
Use of chilli block, where chillis are mixed with cow dung, then dried, and then burnt to deter elephants		x	x
Python fat is mixed with seed before planting to protect the arable fields from predators		x	
Burn mohetola (*Indigofera* sp.) to accelerate sorghum maturity		x	
Use of empty containers to scare the birds and elephants		x	
Use of metal reflectors along the fence to scare the elephants		x	
Collect human urine and pour small bits of it on strategic places around the perimeter fence to scare kudus and jackals (mark territory)		x	
Filling a clear bottle with water and placing at strategic places to scare away jackals		x	
Digging a trench around the field to control pests, e.g., corn cricket	x	x	
Magically protect (Go upa masimo) through seeds	x	x	
Use of whey to control aphids	x		
Eucalyptus for the control of weevils in grains	x		

Potential Climate-Smart Agricultural Practices Identified by Farmers and Extension Workers

When designing CSA implementation strategies at farm level, one must consider adaptation options that are well evaluated and prioritized by local farmers in relation to prominent climatic risks in that location (Khatri-Chhetri et al. (2017); FAO 2012). Table 6 is a compilation of how farmers in the three districts ranked CSA practices according to their preferences. At each district, at extension area

Table 6 Ranking of CSA practices in the Gantsi District and Bobirwa and Boteti sub-districts

No.	CSA Intervention	Ghanzi		Bobirwa		Boteti	
		Charleshill	Nojane	Bobonong	Gobojango	Mosu	Moreomaoto
1	Provision of animal shade						
2	Fodder Production						
3	Vaccination and breeding calendar adjustment						
4	Correct Stocking rate						
5	Feed processing						
6	Use of hardy indigenous breeds						
7	Harvesting and processing of enchroacher plant feed						
8	Ethno veterinary practices						
9	Water harvesting						
10	Pasture re-seeding						
11	Manure application						
12	Rotational grazing						
13	Biogas production						
14	Conservation and utilization of indigenous breeds						
15	On-farm AI of Cattle						
16	Harvesting indigenous plant species for feeds						
17	Use of indigenous poultry breeds						
18	Dip tanks						
19	Mobile laboratory						
20	Holistic pasture management						
21	Solar energy use						
22	Integrated farming						
23	Feedlot weaner production						
24	Small stock AI						
25	Termites chicken feed production						
26	Leather processing						
27	Kraal rotation						

Key	Rating	not rated	High	Medium	Low
	Code				

level, through focus group discussions, farmers came up with a list of CSA practices that existed or was perceived to be of importance for their area. The list was presented to extension workers for validation and in many instances, extension workers added more practices to the list. Only two interventions were highly ranked across the three districts, namely, fodder production and vaccination calendar adjustment. The production of fodder is important because without good-quality animal feed, it would be difficult to have adaptation with mitigation co-benefits of reduced greenhouse gases (Herrero et al. 2013). It has been documented that one of the consequences of climate change is increased incidences of disease (Moonga and Chitambo 2010). Adjusting livestock vaccination calendar is therefore relevant and important to control disease and as seasons seem to have shifted, hence the need to vaccinate timeously. Out of the three districts, Gantsi has the most uniform physiography, for example, its soils are predominantly sandy (arenosols). Furthermore, livestock production is very important, with one of the highest livestock populations in Botswana. This scenario can explain the uniform agreement in ranking of the CSA interventions (Table 7).

Due to centuries of drought exposure, smallholder farming systems in the three districts have developed some level of resilience to natural droughts. Nevertheless, the frequent and severe climate change-induced droughts are beyond the coping capacity of these systems, hence the need to climate proof them through appropriate climate-smart agriculture practices. Effective adoption

Table 7 Implementation plan for fodder production

CSA Technology	Activities	Jan	Feb	Mar	Apr	May	Jun	Jul	Aug	Sep	Oct	Nov	Dec
Fodder production	Land preparation									■	■	■	■
	Buying fertilizers									■	■	■	■
	Buying seeds									■	■	■	■
	Planting									■	■	■	
	Fertilizer application									■	■	■	■
	Weeding	■	■	■									
	Harvesting				■	■							
	Drying				■	■	■						
	Milling and Packaging							■	■				

of climate-smart agriculture requires active participation by farmers not only in identifying constraints to their production but also in developing CSA practices for addressing the identified challenges. This approach requires hybridization of indigenous knowledge with technology. Resilient farming systems lead to sustainable livelihoods, thus, the need for an appreciation of smallholder livelihoods as the anchor on which to build them on. Successful CSA adoption requires co-development of implementation plans by farmers and extension workers.

References

Adger WN (2003) Governing natural resources: institutional adaptation and resilience. In: Berkhout F, Leach M, Scoones I (eds) Negotiating environmental change: new perspectives from social science. Edward Elgar, Cheltenham, pp 193–208

Adger WN, Huq S, Brown K, Conway D, Hulme M (2003) Adaptation to climate change in the developing world. Prog Dev Stud 3(3):179–195

AGRA (2017) Africa agriculture status report: the business of smallholder agriculture in. Sub-Saharan Africa (Issue 5)

Alemaw BR, Chaoka T, Totolo O (2006) Sustainability of rain-fed agriculture in Botswana: a case study in the Pandamatenga plains. Phys Chem Earth Parts A/B/C 31(15):960–966. https://doi.org/10.1016/j.pce.2006.08.009

Botswana Government, Botswana Investment and Trade Center, 2019. Annual Report

Collinson MP (1987) Farming systems research: procedures for technology development. Exp Agric 23:365–386

Cooper et al (2008) Coping better with current climatic variability in the rain-fed farming systems of sub-Saharan Africa: an essential first step in adapting to future climate change? Agric Ecosyst Environ 126(1–2):24–35

Easterling W, Aggarwal P, Batima P, Brander K, Erda L, Howden M, Kirilenko A, Morton J, Soussana J-F, Schmidhuber S, Tubiello F (2007) Food, fibre and forest products. In: Parry ML, Canziani OF, Palutikof JP, van der Linden PJ, Hanson CE (eds) Climate change 2007: impacts, adaptation and vulnerability. Contribution of working group II to the fourth assessment report of the intergovernmental panel on climate change. Cambridge University Press, Cambridge, pp 273–313

Elum ZA, Modise DM, Marr A (2017) Farmer's perception of climate change and responsive strategies in three selected provinces of South Africa. Clim Risk Manag 16:246–257

Engelbrecht FA et al (2015) Projections of rapidly rising surface temperatures over Africa under low mitigation. Environ Res Lett 10(8):085004. https://doi.org/10.1088/1748-9326/10/8/085004

Eriksen S, O'Brien K, Rosentrater L (2008) Climate change in eastern and southern Africa: impacts, vulnerability and adaptation. GECHS Rep 2008(1):27

FAO (2012) Developing a climate-smart agriculture strategy at the country level: lessons from recent experience. Background paper for the second global conference on agriculture food security and climate change, food and agriculture organization of the United Nations, Rome, Italy

Herrero M, Havlík P, Valin H, Notenbaert A, Rufino MC, Thornton PK, . . ., Obersteiner M (2013) Biomass use, production, feed efficiencies, and greenhouse gas emissions from global livestock systems. Proc Natl Acad Sci 110(52):20888–20893

Holzkämper A (2017) Adapting agricultural production systems to climate change – what's the use of models? Agriculture 7:86. https://doi.org/10.3390/agriculture7100086

Intergovernmental Panel on Climate Change (IPCC) (2007) Climate change, impacts, adaptation, and vulnerability. In: Parry ML, Canziani OF, Palutikof JP, Van der Linden PJ, Hanson CE (eds) Working group II to the fourth assessment report of the intergovernmental panel on climate change. Cambridge University Press, Cambridge

IPCC (2001) Climate change 2001: impacts, adaptation, and vulnerability. Cambridge University Press, Cambridge, UK

IPCC (2013) Summary for policymakers. In: Stocker TF, Qin D, Plattner G-K, Tignor M, Allen SK, Boschung J, Nauels A, Xia Y, Bex V, Midgley PM (eds) Climate change 2013: the physical science basis. Contribution of working group I to the fifth assessment report of the Intergovernmental Panel on Climate Change. Cambridge University Press, Cambridge, UK/New York

Khatri-Chhetria A, Aggarwal PK, Joshi PK, Vyas S (2017) Farmers' prioritization of climate-smart agriculture (CSA) technologies. Agric Syst 151:184–191

Krecjie RV, Morgan DV (1970) Determining sample size for research activities. Educ Psychol Meas 30:607–610

Martin G, Martin-Clouaire R, Duru M (2013) Farming system design to feed the changing world. A review. Agron Sustain Dev 33:131–149. https://doi.org/10.1007/s13593-011-0075-4

Maúre G et al (2018) The southern African climate under 1.5°C and 2°C of global warming as simulated by CORDEX regional climate models. Environ Res Lett 13(6):065002. https://doi.org/10.1088/1748-9326/aab190

Mogomotsi PK, Sekelemani A, Mogomotsi GEJ (2020) Climate change adaptation strategies of small-scale farmers in Ngamiland East, Botswana. Clim Change 159:441–460. https://doi.org/10.1007/s10584-019-02645-w

Moonga E, Chitambo H (2010) The role of indigenous knowledge and biodiversity in livestock disease management under climate change. In: 2nd international conference: climate, sustainability and development in semi-arid regions, August, pp 16–20

Morton JF (2007) The impact of climate change on smallholder and subsistence agriculture. PNAS 104(50):19680–19685. https://doi.org/10.1073/pnas.0701855104

Muyambo F, Bahta YT, Jordaan AJ (2017) The role of indigenous knowledge in drought risk reduction: a case of communal farmers in South Africa. Jàmbá: J Disaster Risk Stud 9(1):a420. https://doi.org/10.4102/jamba.v9i1.42

Mwongera C et al (2017) Climate smart agriculture rapid appraisal (CSA-RA): a tool for prioritizing context-specific climate smart agriculture technologies. Agric Syst 151:192–203

Nkemelang T, New M, Zaroug M (2018) Temperature and precipitation extremes under current, 1.5 °C and 2.0 °C global warming above pre-industrial levels over Botswana, and implications for climate change vulnerability. Environ Res Lett 13(6):1–11

Obopile M, Karabo O, Tshipinare BP, Losologolo M, Nkosilathi B, Ngwako S, Batlang U, Mashungwa G, Tselaesele N, Pule-Meulenberg F (2018) Increasing yields of cereals: benefits derived from intercropping with legumes and from the associated bacteria. In: Revermann R,

Krewenka KM, Schmiedel U, Olwoch JM, Helmschrot J, Jurgens N (eds) Climate change and adaptive land management in southern Africa – assessments, changes, challenges and solutions. Biodiversity & Ecology, 6. Klaus Hess Publishers, Gottingen/Windhoek, pp 265–271

Okonya JS, Syndikus K, Kroschel J (2013) Farmers' perception of and coping strategies to climate change: evidence from Six Agro-ecological zones of Uganda. J Agric Sci 5(8):252–263

Pule-Meulenberg F, Obopile M, Chimwamurombe P, Bernard N, Losologolo M, Hurek T, Sarkar A, Batlang U, Ngwako S, Schmiedel U, Nanyeni L, Mashungwa G, Tselaesele N, Reinhold-Hurek B (2018) Diversity of wild herbaceous legumes in Southern Africa, their associated root nodule bacteria and insect pests. In: Revermann R, Krewenka KM, Schmiedel U, Olwoch JM, Helmschrot J, Jurgens N (eds) Climate change and adaptive land management in southern Africa – assessments, changes, challenges and solutions. Biodiversity & ecology, 6. Klaus Hess Publishers, Gottingen/Windhoek, pp 257–264

Thomas DS, Twyman C, Osbahr H, Hewitson B (2007) Adaptation to climate change and variability: farmer responses to intra-seasonal precipitation trends in South Africa. Clim Change 83(3):301–322

Thornton P, Dinesh D, Cramer L, Loboguerrero AM, Campbell B (2018) Agriculture in a changing climate: keeping our cool in the face of the hothouse. Outlook Agric 47(4):283–290

Tschakert P (2007) Views from the vulnerable: understanding climatic and other stressors in the Sahel. Glob Environ Change 17(3-4):381–396

Williams PA, Crespo O, Abu M (2019) Adapting to changing climate through improving adaptive capacity at the local level – the case of smallholder horticultural producers in Ghana. Clim Risk Manag 23:124–135

Winsemius HC, Dutra E, Engelbrecht FA, Archer Van Garderen E, Wetterhall F, Pappenberger F, Werner MGF (2014) The potential value of seasonal forecasts in a changing climate in southern Africa. Hydrol Earth Syst Sci 18(4):1525–1538

World Bank (2007) World development report 2008: agriculture For Development. The World Bank, Washington, DC

World Bank (2010) Africa's infrastructure: a time for transition. The World Bank, Washington, DC

2

Retracing Economic Impact of Climate Change Disasters in Africa: Case Study of Drought Episodes and Adaptation in Kenya

Mary Nthambi and Uche Dickson Ijioma

Contents

Abstract

Valuation studies have shown that drought occurrences have more severe economic impact compared to other natural disasters such as floods. In Kenya, drought has presented complex negative effects on farming communities. The main objective of this chapter is to analyze the economic impacts of drought and

M. Nthambi (✉)
Department of Environmental Economics, Brandenburg University of Technology Cottbus-Senftenberg, Cottbus, Germany

U. D. Ijioma
Department of Raw Material and Natural Resource Management, Brandenburg University of Technology Cottbus-Senftenberg, Cottbus, Germany
e-mail: ijiomuch@b-tu.de; u.d_ijioma@yahoo.com

identify appropriate climate change adaptation measures in Kenya. To achieve this objective, an empirical approach, combined with secondary data mined from World Bank Climate Knowledge Portal and FAOSTAT databases, has been used in three main steps. First, historical links between population size and land degradation, temperature and rainfall changes with drought events were established. Second, economic impacts of drought on selected economic indicators such as quantities of staple food crop, average food value production, number of undernourished people, gross domestic product, agriculture value added growth, and renewable water resources per annum in Kenya were evaluated. Third, different climate change adaptation measures among farmers in Makueni county were identified using focused group discussions and in-depth interviews, for which the use of bottom-up approach was used to elicit responses. Findings from the binary logistic regression model show a statistical relationship between drought events and a selected set of economic indicators. More specifically, drought events have led to increased use of pesticides, reduced access to credit for agriculture and the annual growth of gross domestic product. One of the main recommendations of this chapter is to involve farmers in designing and implementing community-based climate change adaptation measures, with support from other relevant stakeholders.

Keywords

Drought · Agriculture sector · Logistic regression · Climate adaptation · Kenya

Introduction

The agriculture sector in sub-Saharan Africa accounts for about 30% of the gross domestic product (GDP) and 40% exports in the region (Calzadilla et al. 2013). The sector is characterized by rainfed extensive crop and pastoral systems with low yields below potential (Amjath-Babu et al. 2016). The overreliance on rainfed systems has made the region prone to impacts of increasing temperatures, declining precipitation, changes in surface water run-off, and increased carbon dioxide (CO_2) (World Bank 2008). Low precipitation and high temperatures have reduced surface moisture content and increased evapotranspiration rates respectively causing soil degradation. Frequent weather-related disasters such as floods, droughts, tropical cyclones, and landslides have led to deaths and loss of livelihoods (Lumbroso 2017).

Drought is one of the most recurring climate change disasters for countries on the horn of Africa. This extreme weather event has occurred due to lack or inadequate precipitation leading to water shortage for plants, human, and animal consumption (Muller 2014; Maity et al. 2016). Its occurrence has been associated with global climate change, as there exist scientific evidence associating increased human activities with global climate change shocks such as floods and droughts (IPCC 2007). There are four types of droughts: meteorological (shortage in rainfall), hydrological (insufficient water in rivers and reservoirs), agricultural (inadequate soil moisture), and socioeconomic droughts (Maity et al. 2016).

In sub-Saharan Africa, droughts are common and their economic impacts in general have not been adequately studied. The reason behind this could be the complex nature of drought as it presents complex effects to lives on earth (Wang et al. 2014). Drought accounts for about 5% of the natural disasters but the losses that it causes account for about 30% of what is caused by other natural disasters (Wang et al. 2014). It aggravates water scarcity problem by affecting both surface and groundwater resources leading to reduced water supply, quality, and disruption of wetlands (Mishra and Singh 2011). When it persists for months or years, it causes conflicts between pastoral communities due to competition for the diminishing water and pasture resources (Uexkull 2014; Martin et al. 2016).

To demonstrate the economic impacts of drought, this chapter follows the framework by Freire-González et al. (2017). The authors establish many pathways and the drivers that contribute to the economic impacts of drought, which they classify into two categories: primary and secondary. The primary impacts of drought are those that affect economic agents such as industries, households, government, and the environment, while the secondary impacts lead to fires, desertification, and migration of people, as well as animals (Freire-González et al. 2017). Further, drought affects tangible assets such as land and vegetation cover differently from other natural disasters such as earthquakes, storms, or floods that affect buildings, machines, and other assets. In the long term, drought leads to soil degradation, damages to buildings due to soil sagging, as well as the ecosystems due to excessive groundwater abstractions without enough rainfall to replenish it (Freire-González et al. 2017). Drought leads to an increase in temperature causing excessive loss of water through evapotranspiration thus reducing water availability and quality. People are affected differently by drought depending on their geographical location (Freire-González et al. 2017). For example, the impact of drought on people in developing countries is more pronounced due to limited financial capacity to cope with effects such as malnutrition, starvation, increased pests and disease vectors such as malaria and dengue fever-causing mosquitoes (Freire-González et al. 2017), economic burden, and in extreme cases, death. Freire-González et al. (2017) framework focuses on how water scarcity caused by drought affects the households and the environment (Fig. 1).

It is important to focus on drought because it exacerbates the water scarcity situation, coupled with the fact that Africa is one of the driest continents in the world (Mishra 2014). Drought severely affects the water sector in many parts of SSA region, and the impact of the drought on the sector negatively affects the economy.

Some of the effects of drought on the agricultural productivity are also tabulated (Table 1), and the resulting socioeconomic impacts. The analysis gives a broad picture of the effects of drought on the agricultural productivity and the social aspects of the farmers. In some regions of America, the costs relating the effects of drought on the crop and livestock sectors among households have been quantified. For instance, Craft et al. (2017) carried out a study in Kentucky, USA, and reported a reduction in livestocks and hay produced, estimated at a cost of $143.4 due to

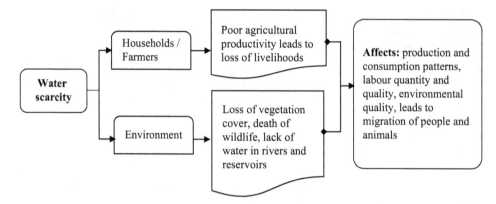

Fig. 1 Impacts of water scarcity caused by drought on households and environment (Adapted from Freire-González et al. 2017)

Table 1 Impacts of drought on agricultural productivity and resulting socioeconomic impacts

Description	Impacts on productivity	Socio-economic effects
Cropping systems	Shift planting and harvesting dates	High market food prices
	Partial or total crop failure	Starvation and malnutrition
	Decreased crop yields	Decrease in farmers' income
Livestock	Decrease in forage from grasslands	Conflicts over pasture
	Increased livestock deaths	Decrease in household income
	Decrease in the quality of meat and milk	From livestock sales
Soil properties and soil health	Insufficient soil moisture for plant growth	Loss of vegetation cover
	Death of soil micro-organisms	Desertification
Hydrological cycle	Reduced water levels in rivers and reservoirs or no water at all	Increase in water prices
	Water scarcity	Conflict over scarce water and pasture resources

water shortage caused by drought. However, the quantitative analysis of the impacts of drought on livestock and crop production in SSA has not yet been estimated due to paucity of available data and research capacity. In the context of SSA, a general qualitative analysis based on previous studies such as Wang et al. (2014), Kusangaya et al. (2014), and Adams et al. (1998) is used in this chapter to draw a close context.

Climate Change Adaptation Measures in the Agriculture Sector of Sub-Saharan Africa

Climate change adaptation is one of the coping strategies supported by the United Nations Framework Convention on Climate Change (UNFCCC) to help developing countries, to reduce the negative effects of climate change (UNFCC 2015;

Deressa et al. 2009). Countries in sub-Saharan Africa express their adaptation needs in a policy document known as the Intended Nationally Determined Contributions (INDCs) (UNFCC 2015). The implementation of the INDCs is supported by several stakeholders such as government institutions, non-governmental organizations (NGOs), donor organizations, the World Bank, and the United Nations Environmental Programs involved in environment and climate change adaptation implementation in SSA region. These organizations have so far supported adaptation strategies in the agriculture sector such as small-scale irrigation of crop in areas where rains have failed, crop and livestock diversification, changes in planting dates, use of soil and water conservation techniques, namely terracing, mulching, and organic manure application (Nthambi et al. 2021; Bryan et al. 2013; Calzadilla et al. 2013; Francisco et al. 2010). Previous research has treated climate change adaptation measures in agriculture as a private good because this only involves individual farmers' adapting to the situation to reduce the local impacts of climate change (Lavoro 2010; Hasson et al. 2010). The effectiveness of these adaptation measures depends on the amount of resources available to support the adaptation process. Adaptation measures can, however, be categorized into two classes: autonomous and anticipatory (public). Autonomous adaptation measures are adjustments to respond to climate variability such as (rainfall) by individual farmers. This could be adopting suitable cropping types, changes in planting dates, and investment in irrigation technologies among other strategies (Calzadilla et al. 2013).

Anticipatory (public) adaptation is a planned measure to respond to climate change which is based on a public policy characterized by acceptability, flexibility, and net benefits (Dinar et al. 2008, c.f. Calzadilla et al. 2013). There are already existing studies from different countries in SSA region that have suggested autonomous adaptation measures (Alemayehu and Bewket (2017); Gebrehiwot and van der Veen (2013); Tessema et al. (2013); Deressa et al. (2011); Amdu (2010)). However, there are limited studies that have sought to address the anticipatory (public) climate change adaptation approach from both a scientific and public policy perspective.

Methodology

Study Area

Kenya is one of the SSA countries in the East African region. It covers an area of 581,309 km^2, which consists of 98.1% land and 1.9% water mass (GoK 2010). The agro-ecological zone distribution map of Kenya is shown in Fig. 2a, and this map identifies four main climate zones including humid, subhumid, semi-arid, and arid regions within the administrative map of Kenya. About 82% of the land mass represents arid and semi-arid areas (ASALs) (Kabubo-Mariara 2009; Rosenzweig et al. 2004), which are naturally prone to drought events. In Kenya, most

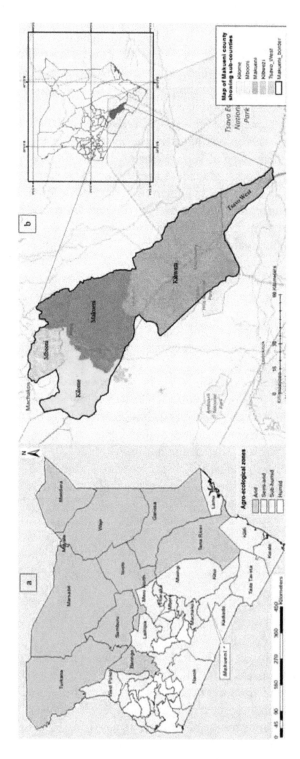

Fig. 2 Map of Kenya showing (**a**) the distribution of agro-ecological zones across the counties and (**b**) the sub-counties in Makueni, the case area in Kenya

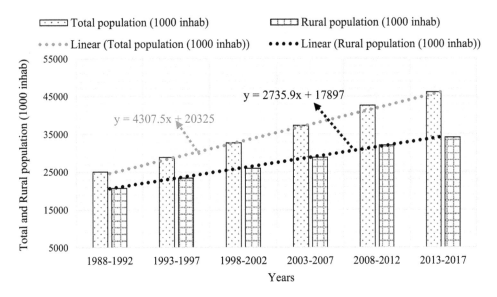

Fig. 3 Total and rural population (1000 inhabitants) between 1988 and 2017 in Kenya (Data source: FAOSTAT 2017)

households found in the ASAL region have limited financial capability to respond to weather shocks such as the severe drought events (Ng'ang'a et al. 2016).

One of the counties in the ASALs region of Kenya in focus is Makueni County. It is situated in the eastern part of Kenya and lies between Latitudes 1° 35′ and 3° 00″ South and Longitudes 37° 10′ E and 38° 30′ E. It covers an area of 8034.7 km² (GoK 2013) I has four sub-counties (Fig. 2b). Households in Makueni County hold a base area of 1.58 hectares which is more in comparison to the national household land holding size of 0.97 hectares. The area experiences a bimodal rainfall pattern of between 300 mm and 1200 mm/annum (GoK 2013). The County's average temperature ranges between 20.2 °C and 35.8 °C (SACRED AFRICA 2011). It is prone to frequent droughts which are caused by inadequate rainfall leading to loss of fodder and food crop productivity. The drought events affect crop farming in the lower areas of the County making the communities living there to take a shift from crop production to animal production (GoK 2013).

One of the main drivers of climate change in Kenya is land degradation resulting from unsustainable land use practices due to a growing population. Land degradation occurs mainly due to the excessive conversion of woodlands into agricultural and settlement areas to cater for the increasing demand for food and shelter among households (Shadeed and Lange 2010). The estimated population in Kenya is approximately 46 million people out of which 34 million people live in rural areas (FAOSTAT 2017). Since the late 1980s, there has been an increasing population trend in the rural areas and the total number of inhabitants in Kenya (Fig. 3). The increase in rural population (Fig. 3) means an increasing dependence on agricultural activities to sustain the daily livelihood of the rural populace.

Muriuki et al. (2011) argue that the population growth in Kenya over the years has led to the massive movement of people from humid and sub-humid areas to arid and semi-arid areas. Resulting land degradation from these migrations has partly contributed to increased temperature and drought frequencies in Kenya since has potential to cause an interruption to the hydrological and ecosystem cycles.

Methods

The selected data mined from databases of the World Bank and the UN FAO websites to evaluate the impact of drought consisted of climate indicators (temperature and rainfall values), population size, average food value, number of undernourished persons, renewable water resources, agriculture value added annual percentage growth, pesticide use, producer prices (US dollars), credit to agriculture (US dollars), and annual % growth of GDP for a period of over 20 years. The time series trend analyses of rainfall, temperature, population size, average food value, number of undernourished persons, renewable water resources, and agriculture value added annual percentage growth were done. Logistic regression model of the economic indicators (pesticide use, producer prices (US dollars), credit to agriculture (US dollars), and annual percentage growth of GDP) datasets were implemented in STATA to demonstrate the relationship between drought events and the selected variables. A binary coding approach was adopted to separate the years when drought events occurred which were coded as 1 and 0 otherwise. The binary logit model estimates the relationship between drought and several other independent variables. The model is applied to variables which are within $(0 \ldots 1)$ and $(-\infty \ldots \infty)$ and is expressed in eqs. 1, 2 and 3 as specified in Bera et al. (2020).

$$c = \beta_0 + \beta_k x_k \tag{1}$$

$$c = log_e [\frac{P}{1-P}] = logit(P) \tag{2}$$

where P represents the probability of a drought event causing an impact to the agriculture sector. c represents the logit(P) and it is related to the independent variables. β_0 denotes the constant value, β_k represents the coeficient estimates for the x factors. x_k represents selected indicators that explain the dependent variable. In terms of the probability that drought events influence the independent variables, the model can be formulated as follows:

$$P = \frac{\exp(\beta_0 + \beta_k x_k)}{(1 + \exp(\beta_0 + \beta_k x_k))} \tag{3}$$

To demonstrate the willingness of smallholder farmers to be involved in an appropriate anticipatory climate change adaptation strategy, focused group discussions were conducted among three population groups in Makueni County. The population consisted of farmer groups, NGOs dealing with agriculture and environmental issues in each of the agro-ecological zones, and government institutions charged with the responsibility of food security and environmental protection. The selection of these three populations was guided by the Agricultural Sector Development Strategy (ASDS) 2009–2020. Six focused group discussions were conducted with six registered farmer groups (a total of 190 farmers; 60 males and 130 females) between February and March 2015. The semi-structured interviews consisted of 17 key informants from the NGOs and government institutions which were used to supplement the information obtained from the smallholder farmers in the focus group discussions. In both cases, farmers and officials of NGOs and government institutions were asked to identify different adaptation measures based on the status quo then.

Results and Discussion

Drought Identification and Historical Trend Analysis

Drought Episodes

The meteorological drought is responsible for hydrological and agricultural droughts types in Kenya. However, a successful integrated water resources management can help prevent the socioeconomic drought caused by the two droughts types. The socioeconomic drought is driven by factors of demand and supply of crop and livestock products (AMS 1997). The meteorological drought events in Kenya recorded were experienced between 1991–2, 1992–3, 1995–6, 1998–2000, 2004, 2006, 2008–9 (Ochieng et al. 2016), 2010–11, 2014–15, and 2016–17 (Reliefweb 2017). These episodes affected the agriculture sector and posed a risk on the food production systems. They caused an imbalance in the production and consumption patterns, which are very important components of food security (Kogan et al. 2013; Li et al. 2009; Tubiello et al. 2007). Previous studies in Kenya show that the agriculture sector is adversely affected by drought events leading to water scarcity, food insecurity, and conflict among pastoral communities (Muriuki et al. 2011). In some cases, people, livestock, and wildlife have lost their lives through starvation (Ombis 2013). The most recent instances occurred on February 10, 2017, and the government of Kenya responded by declaring drought a national disaster. This is because 2.7 million people faced the risk of starvation and about 23 out of a total of its 47 Counties were alarm areas, requiring urgent food and water aid (Reliefweb 2017). The declaration simply means high opportunity costs of other sectoral developments would be lost in pursuing food and cash transfers to the affected people in areas such as the Southeast, coastal lowlands, and the northern parts of the country. Although these impacts of drought are real in Kenya, there are limited scientific documentation to evaluate the impact of drought due to paucity of data. Moreso, there is limited literature which have studied the frequent droughts and linked them to the historical changes in the rainfall and temperature trends in the area.

Rainfall Trends

Over the past decades, scientists have carried out studies demonstrating how rainfall variability and increase in temperature have affected the agriculture sector and/or have a likelihood of causing catastrophes in the future. In East Africa, the annual rainfall amounts have declined considerably since the 1990s thus marking the region with frequent severe droughts between 2004–2005, 2009–2011 (Bloszies and Forman 2015) and 2014–2017 (Reliefweb 2017). The severe droughts have caused failures in rainfed crop production systems leading to food insecurity, famine, and political instability in most parts of northern Kenya, and other countries on the horn of Africa such as Ethiopia, southern Sudan (Bloszies and Forman 2015), and Somalia. Kenya has two main rainy seasons between January to May and October to December and a dry season from June to August. However, since the 1960s the amounts of rainfall received have reduced and are expected to reduce in the coming years. Trend analysis of data from World Bank (2017) database demonstrates a declining trend of precipitation and a high variability of rainfall characterized by extreme highs and lows over the past 55 years (Fig. 4).

A Kolmogorov-Smirnov (K-S) on the rainfall data from 1961 to 2016 indicates a normal distribution of rainfall. However, a Mann-Kendall (MK) statistic test at 95% and 99% indicate significant annual and seasonal variations of rainfall. The changes in seasonal and annual variations represent a reducing rainfall trend (Fig. 4). The trendline in the illustration indicates decreasing rainfall trend (see blue dotted linear) over time. A further decline may be expected if the diagram does not take into account the years of El Niño/Southern Oscillation (ENSO) rains which represent a climate change shock. For instance, the El Niño rains caused massive flooding and

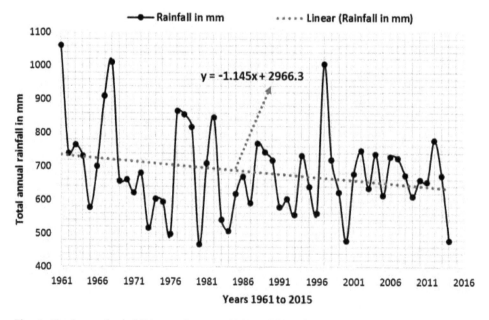

Fig. 4 Total annual rainfall in mm between 1961 and 2014 in Kenya (Data source: World Bank 2017)

mudslides in 1998 (Bloszies and Forman 2015). The severe floods in 1998 caused deaths, extensive damage to infrastructure leaving many households isolated and displaced (FAO 1998). If the years of El Niño rains are excluded from Fig. 4, the rainfall trend would probably be expected to decrease further.

Temperature Trends

Since the 1960s the mean annual temperature in Kenya has increased by 1.0 °C at a mean rate 0.21 °C per decade. The increase is higher between March and May (0.29 °C) and low between June and September (0.19 °C) per decade (World Bank 2017). In the last 55 years, trend analysis of temperature data from the World Bank (2017) has increased by 1.1 °C. The increase in temperature affect the agriculture sector as the soil organic matter is depleted leading to degradation (Kirschbaum 1995). Soil degradation affects crop growth, and causes poor production yield. This is because there is a high correlation between soil organic carbon pool and climate (Kirschbaum 1995). The increase in temperature leads to an increase in humidity in the atmosphere due to high evapotranspiration rates (Tian and Yang 2017). The analysis of the temperature time series of Kenya is illustrated in Fig. 5. The result of the trend in the temperature changes is expected to increase further as shown by the trend (blue dotted linear line) line. The temperature increase in Kenya explains the local warming and drying of the country which has manifested in the form of frequent severe drought events since the 1990s.

Global warming has been associated with increased drought frequencies over the last five decades in Kenya. Yin et al. (2014) argue that the dry to very dry areas have doubled in the world due to effect of rising temperature. In Kenya, it is evident that

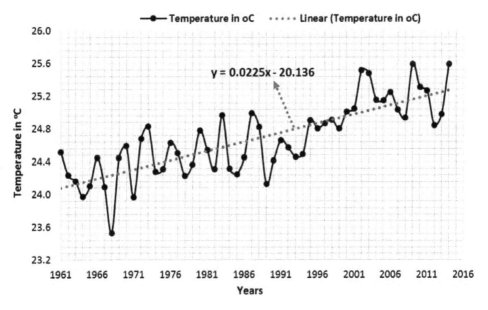

Fig. 5 Mean annual temperature between 1961 and 2014 in Kenya (Data source: World Bank 2017)

the temperature increase has led to further drying in many parts of the arid and semi-arid region, for example, in the south-east, coastal areas, and northern parts of the country. This has increased the cases of land-related conflicts, as viable agricultural crop lands encroach into meadows meant for animals of pastoralists who graze in these areas (Reliefweb 2017).

Evaluation of Economic Impacts of Drought on the Agriculture Sector in Kenya

Impact of Drought on Maize Production

The impact of the global climate change has been linked to the negative effects of economies, governance structures, and communities (Linke et al. 2015). There-fore, many developing economies and poor communities that depend on the agri-cultural produce can be badly affected due to the change. For instance, maize is one of the staple food crop in Kenya and contributes significantly toward food security (MALF 2017). The dwindling trend in its production since the 1990s can be attributed to severe droughts, especially during periods of drought episodes. To illustrate this, the trend in the annual percent change of maize yields corresponding to periods of drought between 1991 and 2014 declined (blue dotted trend line) (Fig. 6). The annual percentage changes in maize production were positive for the years when the drought events of 1991, 1993, 1998, 2008, and 2011 occurred. They became negative in 1992, 1995, 2004, 2010, and 2014 and zero for drought event of 1996.

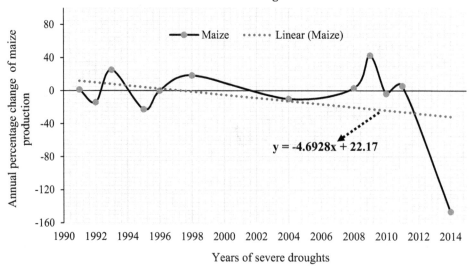

Fig. 6 Annual percentage change in maize production between 1991 and 2014 in Kenya (Data source: FAOSTAT 2017)

During these drought events there were low maize yields and this led to increase in maize prices and higher demand for substitute food types such as rice, wheat, and root crops among others. The increased demand in the substitute food crop for the staple maize, forced the prices of these crop to go higher.

Impact of Drought on the Average Food Value Production

Drought affects the average value of food production expressed in dollars per capita. Average value of food production is a country-wide measure of the absolute economic size of the food production sector (Land Portal 2018). The relationship between rainfalls and average food value revealed that there were drought events between 1991–1998 and 2008–2013. During this period, the average food value in dollars per capita followed a declining trend (Fig. 7). The average food value indicator is estimated over 3 years moving average to reduce the impacts of production errors resulting from complexities that occur due to disparities in major food stocks (Land Portal 2018). The two-person moving average curve (blue dotted curve) in Fig. 7 smoothens the short-term fluctuations in the average food value and shows an increasing or decreasing trend in the long term.

Impact of Drought on the Number of Undernourished People

The agriculture sector in Kenya provides means of livelihood to more than 80% of the population (Faostat 2019). The population of undernourished people between 1990 and 2016 remained relatively high corresponding to the period of drought episodes in the country. These numbers of people are estimated to be in danger of calorie inadequacy and are under hunger threat (World Hunger Education Service 2016). They are thus at a risk of malnourishment. The analysis of the data revealed that the number of

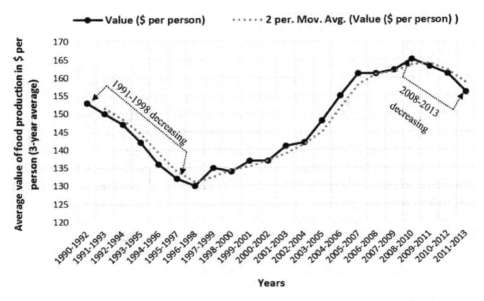

Fig. 7 Average value of food production in dollars per person (3 – year average) in Kenya (Data source: FAOSTAT 2017)

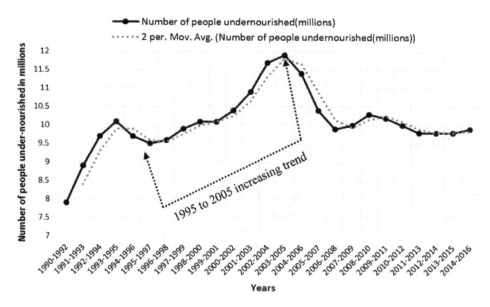

Fig. 8 Number of undernourished people in millions (three-years average) in Kenya (Data source: FAOSTAT 2017)

undernourished people has been fluctuating, but relatively increasing in number, as shown in the trend plot (Fig. 8). For instance, between 1995 and 2005 there is a noticeable increase in the number of people under the threat of hunger in Kenya due to the 11 drought events that occurred between 1995 and 2005. The implication of this fact is that frequent severe drought events can be a threat to the Sustainable Development Goal (SDG 2) of the United Nations on the eradication of hunger and malnourishment because drought limits food availability to the people.

Thus, the number of undernourished people in Kenya have remained high regardless of efforts to meet SDG 2 of ensuring zero (0) hunger. Ensuring zero (0) means ending hunger by achieving food security through access to the right amounts and quality of food from sustainable agricultural practices. To understand the relationship between food and drought, or hunger and water scarcity can be correlated as common phenomena (Loewenberg 2014). This is due to overreliance on rain-fed agriculture which is highly vulnerable to the frequent severe droughts in Kenya. An estimated 75% of the population in the country derive their livelihoods from crop and livestock production (Loewenberg 2014), half of whom survive on less than a dollar a day. People who live on less than a dollar a day to meet their food and water needs are said to be poor (Nthambi et al. 2021; Loewenberg 2014).

Impact of Drought on the Gross Domestic Product (GDP)

The agriculture sector contributes approximately 25.9% to the GDP in employment and in providing food to local communities in Kenya (Ochieng et al. 2016). This means that the impact of drought on the agriculture sector is likely to affect the national GDP. Data from the World Bank (2017) show that agriculture, value-added annual % growth values are negative during the years of drought in Kenya. For example, the annual % growth for the valued added in agriculture 1991 (−1%), 1992

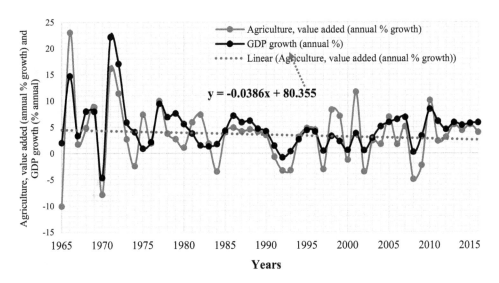

$$y = -0.0386x + 80.355$$

Fig. 9 Agriculture, value added annual % growth versus GDP annual % growth between 1965 and 2016 in Kenya (Data source: World Bank 2017)

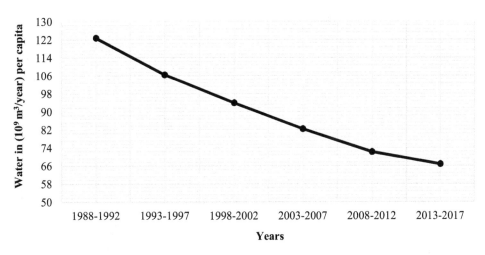

Fig. 10 Total renewable water resources in m³ per capita in Kenya (Data source: FAOSTAT 2017)

(−3%), 1993 (−3%), 2008 (−5%), and 2009 (−2%) (Fig. 9). The agriculture, value-added annual % growth describes the net value output of the agriculture sector obtained by summing up all the outputs minus intermediate inputs. When agriculture, value-added annual % growth curve is compared with that of the GDP % annual growth in Kenya for the period between 1965 and 2016, the curves behave similarly. That is, as agriculture, annual % growth increases, the GDP annual growth % also increases and vice versa (Fig. 9). These trends suggest that the agriculture sector correlates linearly to the GDP. This means that the impact of drought on the agriculture sector will lead to negative value-added annual % growth, which in turn leads to the reduction of the GDP annual growth %.

Impact of Drought on Water Security

Kenya is water-scarce and has only about 3–4% of its land cover composed of wetlands. The country experiences loss of wetlands estimated at 7% per annum due to the expansion of land for crop and livestock production (Böhme et al. 2016). Therefore, severe droughts have led to decline in surface and groundwater recharge. The total renewable water resources per capita per year between 1988 and 2017 have followed a declining trend (Fig. 10). Total renewable water resources are internal and external water resources (both surface water and groundwater) generated through the hydrological cycle (FAO 2016). The internal water resources are generated through endogenous precipitation while external water resources are the resources that enter the country through the transboundary flow from external surface and groundwater resources (FAO 2016). With the total renewable water resource being finite and the population continuing to grow, the water consumption per capita will dwindle. The drought events will have very harsh impact on the population. Hence, alternative water sources and conservation techniques are required to supplement the deficits caused by the drought.

Statistical Analysis of Drought Impact on Selected Economic Indicators

Table 2 shows the descriptive statistics of the dataset obtained from FAOSTAT 2019, consisting of some economic indicators used in the binary logit regression model. These indicators represent independent variables that were hypothesized to be influenced by drought in the agriculture sector. The producer prices represent the average cost of producing a ton of maize – the staple food crop per annum. The annual growth of GDP (%) is the macroeconomic indicator representing the status of the economy in which agriculture, forestry, and fisheries are major contributors. Credit to agriculture represents the average amounts of loans that producers in the agriculture sector, forestry and fisheries, household producers, cooperatives, and agro-businesses have borrowed from private/commercial banking sectors per year (FAOSTAT 2019) to support agricultural production.

Regression results (Table 3) show that pesticide use has a positive relationship with drought events while the coefficient estimates of producer prices, number of undernourished persons, GDP, and credit to agriculture have a negative relationship with drought events.

Table 2 Descriptive statistics of economic indicators influenced by drought from 1991 to 2016

Economic indicators	Mean	Standard deviation
Pesticide use in tons	0.44	0.26
Producer prices (US dollars)	224.38	84.39
Number of under-nourished persons	10.15	0.86
Annual growth in gross domestic product (%)	7.59	12.023
Credit to agriculture (US dollars)	387.69	273.61

Table 3 Summary of the results of the binary logit regression model

Choice (drought/no drought)	Coefficient	Std. Err	z	P > |z|	95% Conf. Interval	
Pesticide use	9.787	2.956	3.31	0.001***	3.993	15.580
Producer price	−0.024	0.0131	−1.85	0.064	−0.050	0.001
Undernourished	−1.467	0.9917	−1.48	0.139	−3.411	0.476
GDP	−0.176	0.0619	−2.85	0.004***	−0.298	−0.055
Credit resources	−0.013	0.0047	−2.73	0.006**	−0.022	−0.004
Constant	−1979.512	11.857	−166.94	0.000	−2002.8	−1956.3
Number of observations	26					
Wald chi^2(5)	158.7					
Prob > chi^2	0.000					
Log likelihood	−12.659					

Pesticide use has a positive coefficient of variation and significant relationship with drought events. Increasing drought episodes means increased use of pesticides to control disease causing pests and vectors.

During drought events, soil moisture reduces the level of pesticide uptake. Moisture-deficient crops attract insects as such crops are unable to produce metabolites that prevent them from being susceptible to pest attacks (Yihdego et al. 2018). The annual growth percent of GDP is negative and statistically significant to drought events. Drought decreases food availability leading to decline in agriculture value added %. During drought events, farmers' ability to borrow money to support agricultural inputs declines as collateral for loans in form of farm assets becomes less available and valuable.

Identified Climate Change Adaptation Measures in Makueni County

Autonomous Climate Change Adaptation Measures

Makueni county has pilot private ponds already in place that are owned by a few households (Fig. 11a, b). Water accumulates in the ponds during the rainy season, and the farmers use it to irrigate the farm and/or fish farming. A standard water pond

Fig. 11 Water ponds for harvesting during rainwater (**a**) dry season and (**b**) filled pond in the rainy season (Photos taken by Mary)

Fig. 12 An example of a household owned water pond for irrigating a kales/vegetable garden showing (**b**) drip and (**c**) furrow irrigation (Photos taken by Mary)

holds a minimum of 200,000 L of water and can be used by an individual farmer to farm an acre of maize and vegetables for approximately three seasons (Fig. 12). However, the disadvantage of this adaptation measure is that most farmers are unable to purchase the polythene sheet used as an inner lining of the water pond because it is costly (KES 70,000 exchange rate @ 103.2270 KES = 1 USD as at June 2015). The plastic layer is used to prevent the harvested rainwater from permeating into the groundwater. However, these water ponds can be a health hazard and that can lead to the drowning of livestock and children if not well guarded, as well as act as a breeding ground for malaria-causing mosquitoes.

Anticipatory Climate Change Adaptation Measures

Anticipatory climate change adaptation coping strategies to climate change are those public adaptation measures provided to the rural communities through the support of organized groups such as the farmer networks/groups, NGOs, and government institutions. These measures are in most cases capital intensive and might require highly skilled technical input to put them in place. A universal set of community-based water projects suggested include boreholes, earth dams, shallow wells, water tanks, subsurface, and sand storage dams. The communities chose a sand storage dam as the best option because of the following reasons; (1) it is the world's cheapest way of providing rainwater to communities in arid and semi-arid areas, (2), a sand storage dam creates an indigenous, reliable, clean water supply up to about 1000 people, (3), water sinks through the sand to the bedrock preventing it from evaporation and pollution as compared to water in shallow wells, and (4) it saves farmers a lot of stress regarding land disputes since it is constructed along seasonal rivers which are owned by the government and (5) sand storage dam reduces the risk of flooding and increases chances of water availability during the dry seasons. A sand dam is constructed across a shallow riverbed such as an ephemeral stream and water is extracted using pumps. An example of sand dam is shown in the following photograph (Fig. 13).

A subsurface sand dam is an underground facility that can be used as an alternative to the surface dams. It has some advantages over the conventional surface

Fig. 13 A sand storage dam construction in Makueni County during the dry season (Photo taken by Mary)

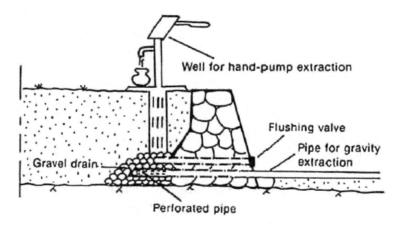

Fig. 14 Cross-section of water extraction from a sand-storage dam (modified from Hanson and Nilsson 1986)

dam because it prevents excessive loss of water through evaporation and does not support the breeding of vectors such as malaria-causing mosquitoes on it. However, the subsurface dam is constructed below the ground level to tap from the flowing groundwater in the natural aquifers. It also has enhanced surface permeability to cause faster infiltration of rainwater to the groundwater during the rainy season. The sand-storage dam stores water in the sand which accumulates naturally behind a dam wall (Onder and Yilmaz 2005). The wall constructed across the seasonal river should be tight and stable to withstand the pressure of running water downstream. The riverbanks of the sand storage dam ought to be protected from erosion on sloppy areas and the dam bottom (Nilsson 1988). According to Hanson and Nilsson (1986), the extraction of the water from the sand storage dam can either be through pumping or gravity. The use of gravity means a drain is placed at the reservoir bottom in the upstream direction of the sand storage dam or a pipe downstream as shown in Fig. 14. The construction of a sand storage dam has four stages: the first stage deals with the capacity of the dam. It is described by the height of the dam wall to be constructed and the quantity of water it is expected to hold.

The second stage involves a careful selection of a proper water pumping mechanism and how to incorporate it into the sand storage dam design. The third step involves the protection of the dam banks through the planting of trees on the sloppy areas both downstream and upstream of the river where the sand storage dam is constructed. The fourth stage involves a careful selection of the donor (management structure) to support the construction of the wall and one that would organize the farmers who will be involved in the construction process. The loss of benefit accruing from the choice of management structure in the area has been fully documented in Nthambi et al. (2021). Farmers should be involved in the four phases and the actual construction process of the selected climate adaptation measure. Other forms of participation can be accounted through contribution of labor time or cash for the purchase of raw materials.

Conclusions

The frequent severe drought events in SSA region have heavily affected the growth of the agriculture sector, forcing crop and livestock farmers to work under very dry conditions in countries such as Kenya. In this chapter, the impacts of drought on the economy of Kenya were discussed, and a possible climate change adaptation measures demonstrated using a case study area in one of the counties. The findings can be summarized in three main ways. First, the analyzed climate indicators revealed frequent drought episodes, which have led to direct and indirect economic impacts on the agriculture sector and the economy. The analysis shows drought events have a positive relationship with pesticide use and negative relationship with credit resources, producer prices, number of undernourished people, and annual percentage growth of gross domestic product (GDP). Again, drought events negatively affect the number of undernourished people as the quantities of crop yields, for example, maize as the staple food crop in Kenya reduces. During the years of drought events, the agriculture value added % growth declines reducing the annual GDP. Producers and households' access to credit resources reduce during drought episodes. The amount of water from renewable water resources also declines due to decreasing rainfall amounts causing water scarcity.

Second, farmers and households can adapt to the negative impacts of drought using either autonomous or anticipatory climate adaptation measures. The major concern is the dryness that comes with drought events, and given the arid and semi-arid nature in most parts of Kenya, rainwater harvesting in sand dams would be the most suitable adaptation measure to reduce the impact of drought. Generally, farmers in the area prefer a sand dam over water ponds or boreholes. However, the lack of financial support hamper farmers' desire to execute this type of anticipatory climate change adaptation measure.

Third, from a methodological perspective, the use of available secondary data to avail important literature required for climate change research in SSA region was promoted. It demonstrates a simple way of relating climate variables with economic indicators using existing qualitative and quantitative approaches. In terms of determining suitable climate change adaptation measures, we consulted stakeholders to suggest the adaptation measures and propose a practical approach to implement them.

Two main recommendations were suggested in this chapter. First, the Kenya National Adaptation Plan (NCCAP) 2015–2030 highlights financing as one of the limitations in the mainstream of climate change adaptation in the water sector. The country is keen on providing adequate water management strategies that can ameliorate the water scarcity problem caused by drought events. We established the involvement of stakeholders to participate in constructing sand dams to harvest rainwater across seasonal rivers. Thus, the NCCAP, 2015–2030 should integrate community participation strategy to provide local materials such as water, sand, and stones or cash when possible, as a way for farmers to partially finance the provision of adaptation measures for sustainable water harvesting techniques. A bottom-top approach to adaptation measure is necessary to understanding farmers' willingness to take part in adaptation projects such as the construction of a sand dam in the SSA region.

The National Drought Management Authority (NDMA) in Kenya, charged with the responsibility to identify risks, end drought emergencies, and ensure adaptation for sustainable livelihoods, should encourage farmers to form community-based organizations. NDMA should also encourage farmers to form farmer networks/groups and register them under the Self-help Association Bill (2015) to provide legal protection to farmers who wish to participate in sand storage dam construction. Membership of community-based organization would encourage collective action efforts that enhance trust among farmers thus strengthening stakeholders' participation in community adaptation projects. Government institutions and non-governmental organizations should work with farmer networks to ensure adaptation measures are implemented in a way that is acceptable to the communities that benefit from them.

References

Adams MR, Hurd BH, Lenhart S, Leary N (1998) Effects of global climate change on agriculture: an interpretative review. Clim Res:19–30

Alemayehu A, Bewket W (2017) Determinants of smallholder farmers' choice of coping and adaptation strategies to climate change and variability in the central highlands of Ethiopia. Environ Dev:77–85

Amdu B (2010) Analysis of farmers' perception and adaptation to climate change and variability: the case of Choke Mountain, east Gojjam., Available via http://docplayer.net/60845560-Analysis-of-farmers-perception-and-adaptation-to-climate-change-and-variability-the-case-of-choke-mountain-east-gojjam.html. Accessed 21 Jan 2018

American Meteorological Society (1997) American meteorological society policy statement: meteorological drought bulletin. Am Meteorological Soc 78:847–849

Amjath-Babu TS, Krupnik TJ, Kaechele H, Aravindakshan S, Sietz D (2016) Transitioning to groundwater irrigated intensified agriculture in sub-Saharan Africa: an indicator-based assessment. Agri Wat Manag 168:125–135

Bera B, Saha S, Bhattacharjee S (2020) Forest cover dynamics (1998 to 2019) and prediction of deforestation probability using binary logistic regression (BLR) model of Silabati watershed, India. Trees, Forests and People 2:100034

Bloszies C, Forman SL (2015) Potential relation between equatorial sea surface temperatures and historic water level variability for Lake Turkana, Kenya. J Hydrol 520:489–501

Böhme B, Becker M, Diekkrüger B, Förch G (2016) How is water availability related to the land use and morphology of an inland valley wetland in Kenya? Phy Chem Earth, Parts A/B/C 93:84–95

Bryan E, Ringler C, Okoba B, Roncoli C, Silvestri S, Herrero M (2013) Adapting agriculture to climate change in Kenya: household strategies and determinants. J Env Manag 114:26–35

Calzadilla A, Zhu T, Rehdanz K, Tol RSJ, Ringler C (2013) Economywide impacts of climate change on agriculture in sub-Saharan Africa. Ecol Econ 93:150–165

Craft KE, Mahmood R, King SA, Goodrich G, Yan J (2017) Droughts of the twentieth and early twenty-first centuries: influences on the production of beef and forage in Kentucky, USA. Sci Total Environ 577:122–135

Deressa T, Hassan RM, Alemu T, Yesuf M, Ringler C (2009) Analyzing the determinants of farmers' choice of adaptation methods and perceptions of climate change in the Nile Basin of Ethiopia. Glob Environ Chang 19:248–255

Deressa TT, Hassan, RM, Ringler C (2011) Perception of and adaptation to climate change by farmers in the Nile basin of Ethiopia. J Agr Sci 149 (1):23–31

Dinar A, Hassan R, Mendelsohn R, Benhin J (2008) Climate change and agriculture in Africa: impact assessment and adaptation strategies. EarthScan London series

FAO (1998) Global information and early warning system on food and agriculture, special report. http://www.fao.org/docrep/004/w7832e/w7832e00.htm. Accessed 10 Jul 2017

FAO (2016) Food and Agriculture Organization of the United Nations: In Statistics. http://www.fao.org/nr/water/aquastat/data/query/index.html?lang=en. Accessed 19 Sep 2017

Faostat (2017) Food and Agriculture Organization of the United Nations: In Statistics http://www.fao.org/faostat/en/ Accessed 25 Jul 2017

Faostat (2019) Credit to agriculture. http://www.fao.org/faostat/en/#data/IC. Accessed 10 Dec 2019

Francisco A, Fredrik C, Maria AN (2010) Farmers' adaptation to climate change. A framed field experiment. Environ for Dev Discussion paper series http://www.rff.org/files/sharepoint/WorkImages/Download/EfD-DP-09-18-REV.pdf. Accessed 20 Sep 2017

Freire-González J, Decker C, Hall JW (2017) Methodological and ideological options. The economic impacts of droughts: a framework for analysis. J. Ecol Econ 132:196–204

Gebrehiwot T, van der Veen A (2013) Climate change vulnerability in Ethiopia: disaggregation of Tigray region. J E Afr Studies 7(4):607–629

GoK (2010) Constitution of Kenya. http://www.kenyalaw.org/lex/actview.xql?actid=Const2010. Accessed 18 Mar 2017

GoK (2013) Makueni County Annual Development plan 2013–14. http://makueni.go.ke/sites/default/files/2013%20Annual%20Development%20Plan.pdf. Accessed 19 Mar 2017

Hanson G, Nilsson A (1986) Ground-water dams for rural-water supplies in developing countries. Groundwater 24(4):497–506

Hasson R, Löfgren Å, Visser M (2010) Climate change in a public goods game: investment decision in mitigation versus adaptation. J. Ecol Econ 70:331–338

IPCC (2007) IPCC Fourth Assessment Report (AR4). https://www.ipcc.ch/publications_and_data/publications_ipcc_fourth_assessment_report_wg1_report_the_physical_science_basis.htm. Accessed 6 Apr 2018

Kabubo-Mariara J (2009) Global warming and livestock husbandry in Kenya: impacts and adaptations. J Ecol Econ 68:1915–1924

Kirschbaum MUF (1995) The temperature dependence of soil organic matter decomposition and the effect of global warming on soil organic C storage. Soil Bio Biochem 27(6):753–760

Kogan F, Adamenko T, Guo W (2013) Global and regional drought dynamics in the climate warming era. Rem Sen Letters 4(4):364–372

Kusangaya S, Warburton ML, Garderen EMV, Jewitt GPG (2014) Impacts of climate change on water resources in southern Africa: a review. Phys Chem Earth 67-69:47–54

Land Portal (2018) Average value of food production (constant I$ per person) (3-year average). https://landportal.org/book/indicator/fao-21011-612. Accessed 10 Aug 2017

Lavoro ND (2010) Adapting and mitigating to climate change: balancing the Choice under uncertainty. Sustainable dev Series: Fondazione Eni Enrico Mattei. http://www.feem.it. Accessed 29 July 2015

Li Y, Ye W, Wang M, Yan X (2009) Climate change and drought: a risk assessment of crop-yield impacts. Clim Res 39(1):31–46

Linke AM, O'Loughlin J, McCabe TJ, Tir T, Witmer FDW (2015) Rainfall variability and violence in rural Kenya: investigating the effects of drought and the role of local institutions with survey data. Glob Environ Chang 34:35–47

Loewenberg S (2014) Breaking the cycle: drought and hunger in Kenya, special report. Lancet 383 (9922):1025–1028

Lumbroso D (2017) How can policymakers in sub-Saharan Africa make early warning systems more effective? The case of Uganda. Int J Disaster Risk Reduction 27:530–540

Maity R, Suman M, Verma NK (2016) Drought prediction using a wavelet-based approach to model the temporal consequences of different types of droughts. J Hydrol 539:417–428

MALF (Ministry of Agriculture, Livestock and Fisheries) (2017) Economic Review of Agriculture (ERA). http://www.kilimo.go.ke/wp-content/uploads/2015/10/Economic-Review-of-Agricul ture_2015-6.pdf. Accessed 27 Mar 2017

Martin R, Linstädter A, Frank K, Müller B (2016) Livelihood security in face of drought – assessing the vulnerability of pastoral households. Environ Model Softw 75:414–423

Mishra AK (2014) Climate change and challenges of water and food security. Int J Sustainable Built Environ 3:153–165

Mishra AK, Singh VP (2011) Drought modelling – a review. J Hydrol 403(1-2):157–175

Muller JCY (2014) Adapting to climate change and addressing drought-learning from the red cross red crescent experiences in the horn of Africa. Weather Climate Extremes 3:31–36

Muriuki G, Seabrook L, Seabrook L, McAlpine C, Jacobson C, Price B, Baxter G (2011) Land cover change under unplanned human settlements: a study of the Chyulu Hills squatters, Kenya. Landsc Urban Plan 99(2):154–165

Ng'ang'a KS, Bulte EH, Giller KE, McIntire JM, Rufino MC (2016) Migration and self-protection against climate change: a case study of Samburu County, Kenya. World Dev 84(2016):55–68

Nilsson A (1988) Groundwater dams for small-scale water supply. IT publications, London

Nthambi M, Markova-Nenova N, Wätzold F (2021) Quantifying Loss of Benefits from Poor Governance of Climate Change Adaptation Projects: A Discrete Choice Experiment with Farmers in Kenya. Ecol Econ 179:106831

Ochieng J, Kirimi L, Mathenge M (2016) Effects of climate variability and change on agricultural production: the case of small-scale farmers in Kenya. NJAS –Wageningen J Life Sci 77:71–78

Ombis GO (2013) Study of climate change impacts using disaster risk reduction interventions to enhance community resilience, case of Makueni County. http://erepository.uonbi.ac.ke/bitstream/ handle/11295/71725/Ombis_Study%20of%20climate%20change%20impacts%20using%20disas ter%20risk%20reduction%20interventions%20to%20enhance%20community%20resilience.% 20Case%20of%20Makueni%20county..pdf?sequence=4. Accessed 15 Apr 2018

Onder H, Yilmaz M (2005) Underground dams. A tool of sustainable development and manage- ment of groundwater resources. Eur Wat 11(12):35–45

Reliefweb (2017) Kenya: drought – 2014–2018. http://reliefweb.int/disaster/dr-2014-000131-ken. Accessed 15 Feb 2017

Rosenzweig C, Strzepek KM, Major DC, Iglesias A, Yates DN, McCluskey A, Hillel D (2004) Water resources for agriculture in a changing climate: international case studies. Global Environ Chang 14:345–360

SACRED AFRICA (2011) Sorghum marketing baseline survey for Nakuru and Makueni Counties for the KAPAP project technical report, Kenya Agricultural Productivity Programme [Unpublished report]

Shadeed S, Lange J (2010) Rainwater harvesting to alleviate water scarcity in dry conditions: a case study in Faria catchment, Palestine. Wat Sci Eng 3(2):32–143

Tessema Y, Aweke C, Endris G (2013) Understanding the process of adaptation to climate change by small-holder farmers: the case of east Hararghe zone, Ethiopia. Agri Food Econ 1(1):13

Tian Q, Yang S (2017) Regional climatic response to global warming: trends in temperature and precipitation in the yellow, Yangtze and Pearl River basins since the 1950s. Quat Int 440:1–11

Tubiello FN, Soussana JF, Howden SM (2007) Crop and pasture response to climate change. Proc Natl Acad Sci USA 104(50):19686–19690

Uexkull NV (2014) Sustained drought, vulnerability and civil conflict in sub-Saharan Africa. Pol Geog 43:16–26

UNFCCC (2015) Adoption of the Paris agreement. UNFCCC, Paris (2015) https://unfccc.int/ resource/docs/2015/cop21/eng/l09r01.pdf. Accessed 22 Sep 2017

Wang Q, Wu J, Lei THB, Wu Z, Liu M, Mo X, Geng GLX, Zhou H, Liu D (2014) Temporal-spatial characteristics of severe drought events and their impact on agriculture on a global scale. Quat Int 349:10–21

World Bank (2008) World development report 2008: agriculture for development. World Bank Washington, DC. https://siteresources.worldbank.org/INTWDR2008/Resources/WDR_00_ book.pdf. Accessed 12 Feb 2018

World Bank (2017) Climate Knowledge Portal for Development Practitioners and Policymakers. retrieved from http://sdwebx.worldbank.org/climateportal/index.cfm?page=downscaled_data_download&menu=historical. Accessed 13 Jun 2017

World Hunger Education Service (2016) World hunger and poverty facts and statistics 2016. https://www.worldhunger.org/2015-world-hunger-and-poverty-facts-and-statistics/. Accessed 10 Oct 2017

Yihdego Y, Salem SH, Muhammed HH (2018) Agricultural pest management policies during drought: case studies in Australia and the state of Palestine. Nat Hazards Rev 20(1):05018010

Yin Y, Zhang X, Lin D, Yu H, Wang J, Shi P (2014) GEPIC-V-R model: a GIS-based tool for regional crop drought risk assessment. Agri Wat Manag 144:107–119

3

Resetting the African Smallholder Farming System: Potentials to Cope with Climate Change

Bernhard Freyer and Jim Bingen

Contents

B. Freyer (✉)
Division of Organic Farming, University of Natural Resources and Life Sciences (BOKU), Vienna, Austria
e-mail: bernhard.freyer@boku.ac.at

J. Bingen
Michigan State University (MSU), East Lansing, MI, USA
e-mail: bingen@msu.edu

Abstract

Agricultural production systems, for example, conservation agriculture, climate smart agriculture, organic agriculture, sustainable landuse management, and others, summarized under the term "sustainable intensification," have been introduced in African countries to increase productivity and to adapt/mitigate CC (CCAM). But the productivity of smallholder farming systems in Africa remains low. High erosion, contaminated water, threatened human health, reduced soil water, and natural resources functionality, that is, ecosystems services, and decreased biodiversity dominate. Low support in the farm environment is also responsible for this situation.

It is hypothesized, based on the huge body of literature on CCAM, that the implementation of already existing arable and plant cultivation methods like crop diversity, alley crops, forage legume-based crop rotations, mulching, organic matter recycling, and reduced tillage intensity will increase CCAM performance and also farm productivity and income. Based on a brief analysis of CCAM relevant arable and plant cultivation methods and agricultural production systems potentials and challenges, this chapter offers guidance for further transforming climate robust African farming systems.

Keywords

Agricultural production systems · Arable and plant cultivation methods · Climate change adaptation and mitigation · Institutional environment · Smallholder farmers practices in Africa

Introduction

In this brief review, the discussion focuses on factors making smallholder farmers vulnerable to CC induced drought and floods and how they are able to cope with CC. The chapter refers to the farms internal management, the relevance of the diverse arable, and plant cultivation methods that are differently organized and highlighted in today's agricultural production systems, which might contribute to adapt/mitigate climate change (CCAM), and how the institutional environment impacts a farmer's capacity to apply CCAM.

This chapter summarizes contributions to CCAM with reference to the diverse mixed farming systems in East Africa. This region is so far of interest as it entails a broad range of African farming systems. Those farming systems with an annual rainfall between approx. 800 mm and 2.000 mm, annual or bi-annual rainfall patterns, and farm sizes between approx. 0.5 and 2.0 ha have been selected. With that the study excludes farming systems from mainly drought regions that however are asking for partly different farming solutions, which would open an additional chapter to discuss. So far, the review will offer a part of the whole picture on farm internal management strategies and relevant environmental (institutional) conditions specifically under East African conditions.

There is evidence that specifically Africans agricultural systems are challenged by climate change (IPCC 2018). To reach the SDG2 targets in the next decade, investment costs to reduce hunger in Africa for the next decade are tremendous (Mason-D'Croz et al. 2019). Such future pictures are based on mainstream agricultural practices, combined with all kinds of single agricultural practices, borrowed from latest discussed agricultural production systems (Gonzalez-Sanchez et al. 2019; Porter et al. 2019; Senyolo et al. 2018; Zougmoré et al. 2018).

The hypothesis, which is framing the analysis, argues that the systemic implementation of already existing arable and plant cultivation methods like crop diversity, alley crops, forage legume-based crop rotations, mulching, organic matter recycling, and reduced tillage intensity will increase CCAM performance and also farm productivity, and as a consequence also the income. The majority of recommendations to cope with CC on smallholder farms commonly involve single methods, such as diversified crop rotations, intercropping, relay cropping, alley cropping and silvo-pastoral systems, compost and manure management, pH regulation, mulching, and low tillage intensity. These methods are summarized as arable and plant cultivation methods, but are also discussed as "agroecological" methods. If properly managed, these methods can increase biodiversity and biomass production; close carbon and nutrient cycles, as well as optimize mineralization processes and increase carbon sequestration; decrease soil erosion; and optimize regulation of the micro- and mesoclimate and the water household. They can also be classified as the backbone of sustainable production, and as such, operate as ecosystems services. The single methods (crop rotation, mulching, . . .) arise in different combinations and partly different "accentuations," in specific agricultural production methods described with the terms "climate smart agriculture, sustainable landuse management, or organic agriculture" and summarized under the term sustainable intensification.

Literature corpus on this topic is huge. Most of the methods are mentioned in the latest report on CC (The IPCC' Special Report on Climate Change and Land. What's in it for Africa? (Dupar 2019 #156). But what is missing in this report is a systemic view and integration of the diverse practices that would make farming systems robust against CC. In other words, to clearly indicate that adoption of farming systems is no longer the adequate reaction on the dramatic situation driven by CC. Instead, there is need for a huge transformation of farming systems as a whole.

For discussing the hypothesis, a selection of reviews and empirical studies seems adequate, using the terminology of arable and plant cultivation, that is, agroecology methods, and selected agricultural production systems (e.g., climate smart agriculture, conservation agriculture and others) within the context of CCAM as key words. Main literature was selected between the years 2010 and 2020 by Google scholar including reviewed papers and grey literature (Majumder 2015), to make visible how far in the last decade, where conservation agriculture, climate smart agriculture, agroecology, or sustainable intensification became prominent in the debate, CCAM was discussed. There are some exceptions, where identified literature informed about earlier relevant sources. Using cropping systems as an example, for all sections with a specific topic, search was defined as follows: Smallholder farming AND cropping systems AND Climate change AND Africa/Review AND Smallholder farming AND cropping systems AND Climate change AND Africa. Following a stepwise saturation in a certain field of knowledge, that is, explicit, reproducible, and leads to minimum bias, further papers that indicate mainly same messages have been excluded (Grant and Booth 2009).

A quantification of applied CCAM is critical due to the fact that there is huge diversity of agroecological conditions and farming practices and that there is no empirical basis available for making such an assessment. Instead, a qualitative assessment of the impact of CCAM methods is applied to primarily understand specific characteristics in their context (see the methodological approach chosen by Dale 2010). For that a set of methods both single arable and plant cultivation methods as well the agricultural production systems have been selected that are well known to be relevant for CCAM.

CCAM Practices and Barriers for Their Implementation

This section reviews the most relevant single methods, their potential to contribute to CCAM, and farm productivity, and the barriers that hinder farmers from implementing them are discussed. Drought adapted varieties, irrigation, and water management technologies are not mentioned specifically. It is assumed that they are part of any CCAM approach.

Crop Diversity and Crop Rotations

In many farming systems, crop diversity in crop rotations is low. Currently, maize is the dominant crop in most crop rotations in Africa. The other land is planted mainly with cereals, grain legumes, and root crops and in some cases intercropped maize and grain legumes (Table 1) (Rusinamhodzi et al. 2012). These one-sided systems, with open soils with low root (< 1 t DM ha-1 a-1) and above ground biomass (< 2 t DM ha-1 a-1), lead to highly vulnerable soils over the whole African continent, continuously loosing fertile soils and the capacity to store water and nutrients. The key crops for moving to higher CCAM relevant ecological functions of cropping systems are forage legumes and forage-grain legumes with high root (> 2 t DM ha-1 a-1) and above ground biomass (> 4 t DM ha-1 a-1) that currently can only be found in some selected regions in pure stands or in combination with maize, known as push and pull system. The latter is a combination of Maize with undersown/intercropped

Table 1 CCAM relevant ecological functions of crops and cropping systems

Ecological functions	Crops							
	Maize[a]	Other cereals[a]	Sorghum[a]	Potatoes	Cassava	Grain legumes[a]	Forage grain legumes[b]	Forage legumes[c]
Soil erosion control	+	++	++	+	+	+	++	++
Carbon sequestration (humus production)	+	+	++	+	+	+	++	+++
Nitrogen fixation	/	/	/	/	/	+	++	+++
Local climate regulation	+	+	++	+	+	+	++	++
Water infiltration/avoidance of water run-off	+	++	++	+	+	+	++	+++
Water holding capacity	+	+	++	+	+	+	++	++

Source: Authors compilation

+ = low; ++ = medium; +++ = high; /=not applicable; − = indicating intermediate position

[a]intercropping of cereals and grain legumes would increase the ecological functions of both crop groups with values between grain legumes and forage grain legumes

[b]crops with high biomass and grain production (e.g., cow pea, grass pea, lablab, mucuna, mung bean)

[c]alfalfa, centrosema, clover, desmodium, and others

desmodium and Napier grass as a trap crop to control stemborer and striga, a chemical ecology-based integrated pest management technology (Khan et al. 2016).

The transformation of cropping systems towards forage legume-based systems can increase the potential of CCAM (Forage legumes > Forage grain legumes > Intercropping systems), through relatively high below (3–6 t DM ha-1 a-1) and above ground biomass (5–15 t DM ha-1 a-1) and nitrogen fixation (50–300 kg N ha-1 a-1) of forage legumes (Kumar et al. 2018), efficiently reducing soil erosion (Sheaffer and Seguin 2003) and with positive impact on following cash crop yields (Traill et al. 2018). Nutrient acquisition and nutrient utilization are species and genotype specific and differ between crop families of grasses and legumes and between legumes (Gómez-Carabalí et al. 2010). Of relevance is the % lignin, C:N ratio, (lignin + polyphenol): N and C-to-N ratios (Thomas and Asakawa 1993). Evidence is also given that this crop group increases animal performance, where similarly the mentioned ratios are of relevance for the digestibility of forage legume feed (Schultze-Kraft et al. 2018).

The implementation of this crop group however is challenging for several reasons: the availability of seed material; the lack of knowledge about the potential impact of plant based nitrogen on the following crop performance; the idea that humans cannot eat forage legumes, and therefore cropping forage is a short-term loss of land for assuring food security, but one that overlooks the contribution of farm yard manure to increased crop yields as well as to the production of milk and meat.

It appears that forage legumes are not offered by either seed breeders, seed wholesalers, advisory services or researchers and policies, with some exceptions of selected species and regions. In the same way, farmers multiplication of seed material is an exception. The diversification of crop rotations with nonmainstream crops is limited because there is no demand by the markets, but also because of traditional food habits. Intercropping and relay cropping – two submethods of crop rotations – likewise can increase yield and income and are rarely implemented.

Mulching Strategies and Green Manure Cover Crops

It is commonly argued that mulch material contributes significantly to the ecological functions and CCAM (Table 2). Already small amounts of mulch material positively impact soil parameters (Mchunu et al. 2011). However, most smallholder farms have no left-over organic matter from their arable fields. All the straw and weeds are consumed by animals after harvest, or burned. As a result, soils are open and this contributes to a further loss of the first most fertile millimeter of humus via wind and water erosion, and leads to a continuous reduction of soil functions. Collecting mulch material outside the farm is for several reasons not a sustainable approach. First, in many regions, biomass is limited, but also requires significant labor, estimated to be 5 to 12 days per acre (Bunch 2017). Furthermore, it does not fit for a community or catchment approach if all farmers would collect biomass, competing also the feed demand of grazing animals, that are mainly far beyond

Table 2 CCAM relevant ecological functions of agroecological practices

Ecological functions	Agroecological methods (arable and plant cultivation methods)					
	Mulching systems	Green manure cover crops	Farm yard manure types[a]	Low tillage systems	Agro-forestry systems	Silvo-pastoral systems
Soil erosion control	+++	+++	+ – +++	+ – ++ +	+++	+++
Carbon sequestration (humus production)	+ – ++	+++	+ – ++	/	+++	+++
Reduction of soil compaction	++	+++	++	+	+++	+++
Nitrogen fixation	/	/–+++	/	/	+++	+++
Local climate regulation	+ – ++	+ – ++	/	+ – ++	+++	+++
Water infiltration, i.e., avoidance of water run of	+++	+++	++	+++	+++	+++
Water holding capacity	++	+++	++	+++	+++	+++
pH-regulation/ increase of nutrient availability	+++	+++	+++	+	+++	+++
Nutrient/carbon recycling			+++	/		
• Spatial (field)					+++	
• Temporal (years)	+++	+++			+++	
• Horizontal (soil layer)	+++	+++			+++	+++

Source: Authors compilation
+ = low; ++ = medium; +++ = high; /=not applicable; − = indicating intermediate position
a including a broad range of types from stable manure up to liquid forms

the carrying capacity of the communities available pasture land (Wall 2017). Further relevant source for mulch material is branches from alley crops with an annual biomass productivity of 2 to 10 t DM ha-1, dependent on the tree species, variety, as well as the planting density and well-managed ratooning (Wilson et al. 1986).

Green manure cover crops are mainly planted in the short rainy season and used also for food and feed purposes. Green manure cover crops (GMCC) include any species of plant, mainly leguminous crops, whether it is a tree, a bush, a climbing vine, a crawler, a grain legume, forage-grain legumes, or close to water a water born plant (Wall 2017). GMCC like Lablab beans (*Dolichos lablab* or *Lablab purpureum*), runner beans (*Phaseolus coccineus*), mucuna (*Mucuna spp.*), ratooned pigeon peas (*Cajanus cajan*), jack beans (*Canavalia ensiformis*), and many other GMCC species like tephrosia (*Tephrosia vogelii* or *T. candida*) produce high amounts of biomass and increase C-sequestration (above and below ground biomass) (Branca et al. 2013; Hobbs and Govaerts 2010). Nitrogen fixation of leguminous crops varies in a broad range of approx. 80 and 250 kg N ha-1 a-1, according to species, variety, soil and climatic conditions, and crop management (Monegat 1991). Assuming a loss of 50% via volatilization and use of grains (Bunch 2017), there are still more than 40 kg ha-1 a-1 N for the following crop with a reference value of 2 t ha-1 a-1 wheat.

Seriously managed, CMCC, combined with minimum soil tillage, can reduce soil erosion by more than 90% (equivalent to a means of 850 kg ha-1organic matter, approximately 50 kg nitrogen ha-1 and 8 kg phosphorus ha-1). GMCC provides nitrogen and offers weed control (Elwell and Stocking 1988), as well as other benefits, including the amelioration soil compacted (Scopel et al. 2004). But GMCC for compost or mulch is only partly a solution because farmers argue that all land should be used for food production, and this hinders them from investing in these crops, which asks for an answer. With forage grain legumes, an alternative exists that can fulfill both ecological functions, but also offer food and feed, and keep relevant parts of the fixed nitrogen in the field. Barriers to their implementation are costs (seeds) and lost harvest of a catch crop income. So, farmers often prefer intercropping of, or rotation with, grain legume crops rather than GMCC (Nandwa et al. 2011).

Farm Yard Manure

Farm yard manure (FYM), which is an un-normed mixture of dung, slurry, straw, and feed residues, is an organic material with many relevant ecological functions (Table 2) (Teenstra et al. 2015). FYM is potentially available on a farm for re-distribution to the fields, if there is any form of zero grazing or animals fenced for certain times in a kraal. Systematically applied, FYM manure can increase biomass production by 100% (Ndambi et al. 2019). In many cases, its impact on yields is comparable with that of medium mineral fertilizer inputs; or the combination of 5 t ha-1 FYM with 30 kg N ha-1 a-1 is able to produce equivalent maize yields in comparison with 200 kg N ha-1 a-1 (Achieng et al. 2010).

Practically each smallholder farm should have some FYM. One ox, one or two cows, and in some cases approx. Five sheep or goats and 10–20 chickens are a classical set of animals for a 1 ha farm. A tropical livestock unit is calculated with 250 kg, which is 50% of the Western standard. Additional feed comes from communal pastures, feed along pathways, and from shrubs. Milk productivity with this feed is rather low (378.6 kg cow a-1 in 2018 in East Africa; http://www.fao.org/faostat/en/#data/QL). As a result, the production of farm yard manure is low as well as their nutrient content. Moreover, this limited manure production is often used for house construction or for cooking purposes. In addition, most farmers do not apply manure management practices, such as roofed animal housing, with water-proof floor or covering manure during storage in order to prevent large nutrient losses during manure storage, increasing greenhouse gas emissions, and reducing the quality of the manure as a fertilizer and its positive impacts on ecological factors, that is, contributions to CCAM (Ndambi et al. 2019).

To conclude, farm yard manure contribution in current smallholder systems to strengthen CCAM or productivity is seriously low or nonexistent. Hurdles for an adequate collection and use of FYM are lack of knowledge in general about the value of this organic matter, techniques for cut and carry feed, investment for zero grazing units, that is, kraal fencing, and techniques for the adequate distribution of the manure in the field. Furthermore, it seems neither in the focus of advisory services nor of research or policy agendas (ibid).

Soil Tillage

Soil tillage functions include, among others, to loosen the soil, to provide technical aid for in depth growth of roots, to homogeneously mix organic matter with the soil, and to prepare seed beds.

Current partly excessive soil tillage systems often lead to soil compaction, soil aggregate break down, soil dust, and rapidly decomposing organic matter. High rates of topsoil loss contribute to downstream sedimentation and degradation of local and regional water bodies (Tully et al. 2015). While soil compaction reduces the seedbed quality and thus hinders the germination of seeds, soil dust gets irreversibly lost via water and wind erosion. Thus, the current type of so-called plowing tillage systems in combination with maize dominated cropping systems weakens several ecological functions relevant for CCAM.

In general soil tillage systems follow a continuum from conventional inversion moldboard plow up to disc opener no-till planter into dead residue (Reicosky 2015). Alternative soil tillage systems have been developed over the years to reduce tillage intensity, including stubble mulching, to reduce or eliminate tillage, and to retain plant residue on the soil surface to alleviate wind and water erosion. These practices led to what became known as conservation tillage, culminating in no-till systems that avoid any soil disturbance. No-tillage, reduced-tillage, mulch-tillage, and strip-tillage are some of the diverse CA tillage approaches.

The combination of green mulch cover crops (GMCC) is expected to significantly increase ecological functions and contributions towards CCAM (Table 2). GMCCs can also control weeds by covering, that is, suppressing upcoming weeds. Potential yields through combining mulching and low soil tillage exceed 100%. But it must be remembered that the quantity of additional organic carbon in the soil under no-till is relatively small. The main effects are relevant for other ecological functions and indirectly contribute to CCAM. Over the years it also has become evident, that after a period of low tillage interventions, this system must be interrupted for 1 year with a more intense tillage approach, in order to loosen natural soil compaction effects in the below ground, as well as to regulate the weed pressure (Powlson et al. 2014).

Soil tillage systems technologies with reduced intensity are only available in some regions and the optimal approach which depends on soil type, rainfall, share of skeleton, precrop and the following crop, as well as above ground biomass is still not identified for many agroecological situations. If not managed properly, nonadequate technique applied, and rotations without forage legumes, as well as limited or lack of mulch material, there is high risk of weed pressure, reduced crop yields. If weeds take over farmers then tend to misinterpret the approach increasing the application of fertilizer inputs and herbicides with negative environmental impact (Giller et al. 2011). A certain number of weeds however can provide also positive ecological functions, as discussed in OA (Hillocks 1998). Missing finances for the technical equipment and knowledge gaps might be other reasons for low adaptation rates of reduced tillage systems in Africa.

Agroforestry

The multifunctions of the diverse agroforestry systems are highly relevant for CCAM (Lasco et al. 2014), but also serve as forage or for other diverse purposes (fuel, construction, apiculture, food, forage), and provide positive contributions to reduce concentration of carbon dioxide (CO_2) and other greenhouse gases (GHGs) (Awazi and Tchamba 2019) and to regulate microclimate (Mbow et al. 2014). Specifically, alley-crop-systems contribute to multiple ecological functions (Table 2). They contribute to a reduction of inorganic inputs (Wilkins 2008), to pest and disease control (Murthy et al. 2013), biodiversity (Murthy et al. 2016), and regulation of microclimate (Schroth et al. 1995). What makes them specifically attractive is their use of subsoil layer nutrients and water resources and photosynthesis above the main crops, well adapted to soil and climate conditions, simulating partly the original vegetation cover before once turned the land from natural forest to arable land.

Alley branches can be used directly as mulch, or as a protein feed for animals, filling the protein gap and leading to higher animal manure production. Trials of alley crops with rows all 4–8 m document the high positive impact on cash crop growth. A weakness in the majority of trials are monocropping alleys with an increased risk of pest and diseases, while diverse alley crops provide more resistance, multiple functions for arable pest and disease control, and a diversification of forage qualities and adaptability towards CC (e.g., *Faidherbia spp.*).

On smallholder farms, their integration is challenged if the land is scattered. Otherwise, already the limited area of the homestead would offer space for implementation. Biomass productive hedges are often not in place and also the technique of ratooning, for example, in case of pigeon pea, and other management practices are missing and are limited in biomass production with all its positive impact on CCAM. Further hurdles for smallholder farmers to integrate a diverse cropping scheme with alley crops are the limited availability of seedlings, lack of knowledge how to prune shrubs and to integrate into feed ratios, as well as investment costs and workload for their management.

(Silvo-) Pastoral Systems

Pastoral systems are usually the backbone of animal systems. By nature, grassland is rich in root biomass and humus content is seriously higher than in arable systems with permanent soil tillage. However, in most African countries, pastures are overused, sward management is low, erosion is high, and productivity is not high enough to feed animals with energy and protein rich crops. Irreversible loss of land via erosive processes/land-slides is point on the agenda explaining the critical situation (Wynants et al. 2019). Land distribution and ownership patterns coupled with continuous fragmentations are further drivers of the current system and are challenging farmer towards sustainable grassland management (Tesfa and Mekuriaw 2014). Provision towards ecological functions, that is, contributions towards CCAM, is low if not negative.

Limited availability of seeds, lack of labor for establishing pasture systems, the cost of fencing and knowledge gaps, as well as missing management and knowledge at community level, as well as missing management at the community level, hinder and limit the production of pasture biomass and thus also their CCAM.

As an alternative silvopastoral system, a combination of pastures with leguminous shrubs and trees provide relevant contributions to ecological functions towards CCAM (Table 2) and an enormous increase of biomass production (Sarvade et al. 2019). The forage legume trees cover a high share of the protein gap limiting the animal production. Forage production can be seriously increased with a strong impact on the level of fattening and milk production and less GHG emissions per animal unit (Broom et al. 2013). For that grassland management needs to reorganize, including regulations for use, awareness on traditions, and fenced or otherwise controlled areas (Ochieng and Waiswa 2019).

Impact of Arable and Plant Cultivation Methods on pH Regulation

The pH of soils with substantial ecological functions and specifically high impact on biomass production is influenced by several of the arable and plant cultivation methods and therefore is summarized in this subsection. The pH of many East African soils is below 5.0, virtually no available phosphorus and toxic levels of aluminum (Bunch 2017) and therefore out of the optimum for a productive crop growth and biomass production, which is between 5.8 and 6.5, while crop specific in a broader range. Mineralization processes, the optimal living conditions for microorganisms responsible for many soil functions, and growth conditions for crops are limited with low pH, leading to lower crop yields, that is, less positive impact on CCAM. Under acid soil conditions, phosphorous is tied up in a short time and does not provide more than 0.5% soil phosphorous in forms that are available to plants. The general low P content of African soils increases the dilemma.

Mulch and compost material mainly from grass provide a large C/N and therefore will not feed the crops. A high share of legumes instead offers a C/N which boosts the growth of microorganism that are functioning as storage but also deliver nutrients towards crops through the decomposition of mulch material. Phosphorous becomes available as a slow floating source. Farm yard manure (approx. pH 7), slurry (approx. pH 8), and compost (approx. > 7 pH) with high pH values would positively influence nutrient availability and thus crop productivity. Current systems are lacking these effects due to the low availability of manure, slurry, and compost, which at least causes the inefficient landuse and low productivity of crop and animal production. Also, alley crops have the potential to increase the pH via litter fall, through re-transporting leached Ca, however depends on tree species, and the presence of a subsoil of suitable quality, that is, clay enriched and with high Ca saturation (Vanlauwe et al. 2005).

Using lime is an option, so far locally available and transportable. But liming runs the risk of increased mineralization, that is, humus decomposition beyond what is needed, and there are cost factors.

Cattle, Dairy, Sheep, and Goat Systems

Ruminants play an intermediate position between crops and manure. Current feeding ratios are often of low quality. Overgrazed grassland, lack of any forage crops in the crop rotation, alley crops, and hybrid grasses (*Pennisetum spp.*; *Brachiaria spp.*) are the explaining factors. Specifically, protein deficits explain also the low performance in dairy production. Intense use of straw and stubble after crop harvests – a main forage source – reduces the humus level of soils, specifically if organic manure is not recirculated to the fields, which is in manure farms the practice and is increasing soil erosion. Without a forage management including access to water, positive outcomes of breeding programs cannot be transferred into practice. To add cereals or grain legumes in the feed ratios compete directly with human food and cannot be recommended as a key strategy, except the use of leftovers.

Ruminant demand for forage legumes, hybrid grasses (Ghimire et al. 2015), forage trees/alley crop branches (Franzel et al. 2014), and silvopastoral systems fit with the CCAM strategies, due to the fact that they all contribute to an increase of biomass production, farm internal nitrogen supply, and milk productivity per land unit. As a result, the organic manure nutrient content and quantities increase, which can serve as a valuable fertilizer for crops and for supporting soil functions.

Agricultural Production Systems Contributions to CCAM

This section first discusses diverse agricultural production methods and how they relate to each other in order to identify communalities and differences. As seen in the description of single CCAM relevant methods, they often refer to CA and OF. Table 3 informs about differences and similarities within an African context.

Sustainable Intensification – Some Terminological Clarifications

Most prominent agricultural production methods in recent years can be summarized as "Sustainable intensification (SI)," which is defined as meeting the growing demand for agricultural production while conserving land and other resources, "from the same area of land while reducing the negative environmental impacts

Table 3 Assessment of CCAM relevant methods in CA and OA

	Agricultural production methods	
	Conservation agriculture	Organic agriculture
CCAM relevant methods		
Crop rotation	++	+++
Crop rotation	++	+++
• Forage legumes	++ – +++	+++
• Relay cropping	+++	+++
• Intercropping	+++	+++
• Green manure	+++	+++
Farmyard manure	+++	+++
Compost	+++	+++
Mulching	+++	+++
Minimal tillage	+++	+ – ++
Hybrid grass	++	++
Alleys/hedges	+++	+++
Silvopast. Systems	++	+++
pH regulation	+++	+++
Biodiversity[a]	+	+++
Other methods		
Nitrogen fertilizer	a	e
P, K mineral fertilizer	a	a (accepted are low soluble fertilizers)
Herbicides	a	e
Pesticides	a	e

Source: Authors compilation
+ = low; ++ = medium; +++ = high; /=not applicable; – = indicating intermediate position;
0 = not explicitly mentioned; a = accepted; e = excluded
[a]via biotope diversity

and at the same time increasing contributions to natural capital and the flow of environmental services" (Khan et al. 2017). This term also accounts for the human condition, nutrition, and social equity. In their review, Smith et al. (2017) conclude, using a broad range of socio-, cultural, and socio-economic measures, that most current agricultural practices fall into this category. SI does not explicitly exclude any farming practice or requires an obligatory production framework, as, for example, the case for conservation agriculture with minimum tillage and mulching, or organic farming that excludes some farming and processing inputs. Mahon et al. (2017) conclude that SI is an oxymoron, underpinned by a productivist agenda, and that it is lacking a clear rationale. In their review, Xie et al. (2019) identify the reduction of yield gap as the main target of SI. But the intense use of mineral fertilizers is also not excluded (Holden 2018). Yet others, like Mdee et al. (2019), understand SI to be based on agroecological methods, or to classify integrated pest management as a core activity (Pretty and Bharucha 2015).

Definitions of CA do not explicitly exclude mineral fertilizers, herbicide and pesticide applications, but keep them at an option. Consequently, Vanlauwe et al.

(2014b) summarize three conceptual pathways for intensification paradigms: (i) integrating soil fertility management ending up in conservation agriculture, (ii) push and pull systems (Khan et al. 2008b), and (iii) evergreen agriculture (agroforestry-based systems) (Garrity 2017). All of these systems contribute to high soil health status and productivity through somewhat different pathways. In contrast to the overall definition of SI, which includes overlaps of sustainable and ecological intensification (Wezel et al. 2015), these pathways are classified as being only part of ecological intensification and thus only cover partially what is discussed under the term SI.

Another concept, Climate Smart Agriculture (CSA), is defined by the FAO as "agriculture that sustainably increases productivity, resilience (adaptation), reduces/removes greenhouse gases (mitigation), and enhances achievement of national food security and poverty reduction" (Nyasimi et al. 2014). CSA refers to food security and rural livelihood improvement through agricultural concepts that facilitate climate change adaptation and provide mitigation benefits (Scherr et al. 2012). Arable and plant cultivation methods applied refer to many of the field-based and farm-based sustainable agricultural land management practices identified as part of SI. As Zougmoré et al. (2016) suggest in their analysis of CCAM relevant approaches, CSA is a farming method like CA.

"Sustainable land (use) management (SLM)" as the last approach, which should be mentioned in this brief analysis, is a compilation of methods that summarizes many of the activities already introduced above. Gurtner et al. (2011) define SLM as "the adoption of land use systems that, through appropriate management practices, enables land users to maximize the economic and social benefits from the land whilst maintaining or enhancing the ecological support functions of the land resources." Arable and plant cultivation methods of SLM include soil fertility and crop management, soil erosion control measures, organic fertilization, minimum soil disturbance, and incorporation of residues, terraces, water harvesting and conservation, and agroforestry, grazing and forest management (Branca et al. 2013; Cordingley et al. 2015). But scholars highlight different specific methods. Dale (2010) defines SLM mainly as a technology approach to cope with erosion via mainly technical interventions and less one defined by crop-based systems.

Conservation Agriculture

Conservation agriculture (CA) integrates three principles (FAO) (www.fao.org/ag/ca): (i) avoiding or minimizing mechanical soil disturbance; (ii) enhancing and maintaining a permanent mulch cover with organic matter on the soil surface; and (iii) diversifying species (Kassam et al. 2017). Today CA is associated with three other agricultural production systems.

Push and pull system discussed as a core subsystem of CA, but also described as a climate smart technology that combines maize with undersown forage legumes and Napier grass, and with a minimum tillage approach (Midega et al. 2015), fulfills many of the ecological functions that contribute importantly to CCAM (Table 3) and enables sustainable cereal crop and livestock production intensification through ecosystem or agroecological approaches based on natural biological processes (Khan et al. 2014). In the face of CC, new drought-tolerant trap (e.g., *Brachiaria* cv. *Mulato*) and intercrop (drought-tolerant species of desmodium, e.g., *D. intortum*) plants have been studied to further develop the push and pull system (Khan et al. 2017). Another approach is Evergreen Conservation Agriculture, a subtype of CA, which focuses specifically on the integration of trees into crop and livestock production systems (Garrity 2017). Close to this system are agroforestry systems that include inter-planting of a broad range of leguminous shrubs (Pratt et al. 2002). For erosion control in combination with other conservation measures, for example, dense hedges of vetiver grass on contours are recommended (Bunderson et al. 2015).

The positive impacts of CA on yields are well documented (Mkonda and He 2017), with yield increases compared to traditional methods, for example, from 1 t ha-1 to 3.5 t ha-1 per cropping season (Khan et al. 2006) as well as increased milk production (Khan et al. 2008a; Khan et al. 2008b; Midega et al. 2014). As a result, CA is shown to contribute significantly to higher economic returns for the farmer (higher returns to both land and labor than conventional farmers' practices), thereby allowing reduced, and sometimes none, use of nitrogen fertilizers, herbicides, and pesticides. CA contributions towards carbon sequestration further underline their high contribution to CCAM (Gonzalez-Sanchez et al. 2019; Sommer et al. 2014).

Organic Farming

Organic farming (OF), as described by internationally accepted guidelines, is designed to be followed for farmers who are approved by certification processes and want to market their products as organic. OF also embodies a set of ethical principles that serve to orient all the subsystems of this agricultural production system. OF relies on (agro-) ecological processes, biodiversity at all levels (species, genotype, habitat, landscape...), and nutrient and carbon cycles, and high contributions to reduce agricultural GHG emissions (Scialabba and Müller-Lindenlauf 2010). All CCAM methods are recommended (see also FAO 2011), while other methods with often critical impact on the environment, but also for CC are limited (Table 3).

A key characteristic of organic farming is the focus on soil fertility, as a result of diversified crop rotations with forage legumes, catch crop and intercropping, and organic manure, which is the basis of a healthy and sustainable production (Meyer 2010). These diversified crop rotations largely replace the functions of excluded inputs like soluble mineral fertilizers, herbicides, and pesticides (Freyer 2019). Forage legume-based crop rotations, agroforestry, and nutrient and carbon recirculation are of high relevance. From a systems perspective, all of these methods must be implemented to enable a productive system. Tillage systems follow the idea to avoid the mix of fertile above ground soil layers with less fertile in lower layers. How far tillage intensity can be reduced depends on precrop – following crop relations, produced above and below ground biomass, organic manure type and amount, and the demand for weed control or to initiate mineralization processes, therefore cannot be generalized (Koepke 2003). At this point the difference between OA and CA becomes relevant which is the exclusion of many external inputs in the OA system. Functions of these excluded inputs must be partly taken over by the tillage system.

Other characteristics of OA are the inclusion of traditional knowledge, material, and methods (Müller and Davis 2009). Also, local seed collection and use is an intrinsic part of organic agriculture (Forster et al. 2012). Such seeds might be an advantage to adapt under certain climatic conditions and contribute to maintaining agro-biodiversity.

Africa hosts the majority of organic farms worldwide, with almost 2.1 million hectares of certified organic agricultural land (2017), and at least 815,000 producers (Willer and Lernoud 2019). But what is currently missing is an evaluation of the practices in these organic farming systems. Annual farm certification only partially informs the extent and quality of CCAM applications.

Moreover, research lacks a convincing organic systems approach. If trials include amounts of organic fertilizers that are going beyond the production capacity or what is currently known on that in practice, crop rotation principles are ignored, and alley cropping is not integrated (Adamtey et al. 2016; Musyoka et al. 2017). Such trials are of low relevance for identifying the real potential of OA to CCAM. This limited availability of research trials hinders a more in-depth assessment of the organic approach. Specifically, in comparison to CA, the gap is huge.

In African countries, OA is often equated with traditional practices – whatever they are – less mechanized, and minimum use of chemical fertilizers and pesticides. The low input systems neither fulfill the organic principles of health, ecology, fairness, and care, are not in line organic guidelines, nor would survive any organic farming certification process (Nalubwama et al. 2011).

Discussion

This section summarizes and provides an orientation to the interdependence of CCAM relevant methods. It informs the challenges of farmers to adopt and refers to farmers' perspectives, agroecological, social, livelihoods, and the surrounding conditions for implementing CCAM, as well as how to interpret communalities and differences between CA and OA.

The Logic of CCAM Practices

The interdependence of CCAM relevant methods is a key to understand their efficiency, success, or failure, not only for CC, but also their impact on productivity. Methods can be classified into two levels:

- First level: Methods that provide biomass and nitrogen: Legume based Agroforestry systems, silvopastoral systems, forage legume-based/green manure cover

crop, relay crop and intercrop-based crop rotation systems, and perennial productive high breed grasses.
- Second level: Methods that depend on the productivity and management of first
  level methods: mulching, manure and compost production, minimum tillage, pH
  regulation, biodiversity for pest and disease regulation.

Of course, isolating the yield effects of individual practices is complicated
(Branca et al. 2013), but without optimal management of first level methods, second
level methods cannot be activated or are weak in their performance. In other words,
if methods in the first level are not well established, second level methods cannot be
put in place. In organic farming, the application of CCAM relevant methods is
obligatory to enable productivity with limited input from outside. Exclusion of one
method would weaken the whole system. Of high relevance is the management
quality of applied methods which allow that, even with very low amounts of
residues, increases in the productivity with respect to conventional farming practices
(Sommer et al. 2012). This fact is often overlooked. OA is very clear on this point –
there are no half measures – if so, the system would fail both in CCAM and in
productivity.

Barriers for Adoption CCAM from a Production Perspective

In recent decades, extensive efforts to promote CA, CSA, or SLM have been
undertaken. Despite extensive past efforts by governmental, nongovernmental, and
research organizations to promote SLM to smallholder farmers in SSA, adoption
remains low (Cordingley et al. 2015). What are the barriers to establish these
methods from a production perspective?

- Crop specific: forage legume seeds, agroforestry seedlings and tree management,
  hybrid grass seeds, pasture seeds, weed control.
- Technically specific: soil tillage and seed technology, technology for cut and carry
  systems, stable construction, water management, storage, transportation and
  distribution of organic manure and compost.

Challenges to implement CCAM approaches are obviously not essentially different between CA and OA. The difference lies in the accessibility of nitrogen fertilizer,
herbicides (Wall 2017), and pesticides that all can be applied as "emergency"
methods in CA, while not in OA, These inputs provide CA farmers a set of methods
that can be used when the crops are in a critical stage (lack of nitrogen, high weed
pressure, expected outbreak of pest and diseases).

The differences between CA and OA can be identified largely in the so-called
"other methods" (Table 3). Neither soluble mineral fertilizers nor herbicides and
pesticides are allowed in either to bridge critical weather conditions, weed, or pest
and disease development. Nitrogen fertilizer is the most relevant input with high
impact on GHG emissions.

The application of these inputs is also controversially discussed in CA. While some argue that the three CA principles are sufficient to guide CA practices (Sommer et al. 2014; Wall 2017), Vanlauwe et al. (2014) identified a fourth principle, the application of mineral fertilizers. Many farmers believe that CA cannot be undertaken without specific inputs and tools, a message commonly conveyed to them by extension staff (e.g., hybrid seed, fertilizers, herbicides, knapsack sprayers, jab planters,...) (Bunderson et al. 2015). For farmers, the exclusion of "other methods" would limit the acceptance of CA and its scale on farms. Farmers also fear an increased abundance of termites and earthworms as harmful, whereas in fact they can have beneficial effects on the soil and crops (ibid). But there are also voices from the organic side that agriculture in Africa without herbicides and synthetic fertilizers is not possible and vote for their integration into the organic guidelines (Lotter 2015). The result would be simply CA. Other surveys report that farmers' perceptions indicate that CA without external inputs, such as fertilizers, herbicides, pesticides, or compost, can improve yields relative to conventional agriculture. This argument is supported by the observations of scientists that a large proportion of soils in sub-Saharan Africa can be described as nonresponsive (Tittonell and Giller 2013), in which mineral fertilizer applications do not result in higher crop yields, while already the application of the three CA principles led to serious higher yields in comparison to the conventional input driven approach (Ngwira et al. 2013; Thierfelder and Wall 2010). Also, effective weed management is possible without the use of herbicides, since all three CA practices contribute to integrated weed control, also experienced in organic farming, if crop rotation and weed management are organized following latest knowledge. These observations are a clear reference about the OA system that also builds besides other methods on the three CA principles. These experiences are also showing that subsidizing of inputs is not a necessary precondition to CA use and uptake (Bunderson et al. 2015; Lalani et al. 2017).

According to FAO (2012), 80% of all farms in Africa have an agricultural area of <2 ha. These fragmented small land holdings preclude the effectiveness of large machinery in fieldwork, specifically for cut and carry systems, mulch-seeding systems/zero-tillage direct seeders, or the distribution of farm yard manure or mulch from outside the field, or from alleys (Sims et al. 2017; Sommer et al. 2014). In contrast, Lalani et al. (2017) observed that manual forms of CA can be attractive for farmers, particularly those with very small plots of land (0.5 ha or less), observing that mulching systems can reduce weed pressure.

Labor for biomass management is discussed as a barrier to the adoption of some CA methods, while others like seed varieties that are drought tolerant and early maturing are classified as the most suited technologies for smallholder farmers to respond react to CC (Senyolo et al. 2018). To define the CA approach only with drought resistant seed material, optimized irrigation and mineral fertilizers fail the overall idea, which could be also titled with "CA by default" or a conventionalized CA; of course it is less complex to apply a hand full of fertilizers and a herbicide instead of developing diversified crop rotation, mulching, and agroforestry systems.

A further barrier of CA/OA systems with the CCAM approaches is that they cannot be put in place in one season. Often the starting points for transforming the system are poor soil fertility, land degradation, pests, and erratic rainfall that are challenging the adoption (Sietz and Van Dijk 2015). It is therefore not only about implementing new farming methods, but also to "repair" the damaged natural system. There is a conversion period of 2 to 3 years as it is also for OA, where farmers primarily have to invest in the transformation of the system. This requires a period of labor and financial investment, and patience by farmers until trade-offs of the new system become visible in an increase of productivity (Lalani et al. 2017).

In addition to these relevant hurdles, it is to keep in mind that weed and pest and disease control are a challenge for all farmers, irrespective of the CCAM methods, and not specific for CA/OA, like diversified crop rotations, in comparison to maize monocropping (Thierfelder et al. 2013; Thierfelder and Wall 2010); however, the high increase of pest and diseases provoked by monocropping and biodiversity poor landscapes underlines that farmers on the long run have no other choice than to modify their system, independent from CC.

Markets, Policies, and Education for a Diversified Production

The implementation process of CCAM methods still lacks a clear conceptual understanding (Partey et al. 2018). The poor understanding of CA/OA, its CCAM methods, and suspected contradictions over what it means reflect the absence of an overall educational training program for teachers, extension officers, and students (Bunderson et al. 2015). Field trials should not lack of first level CCAM approaches (see the example of (Thierfelder et al. 2013, 2016); without training programs for farmers, chances for successful adoption of CCAM will continue slim (Erenstein et al. 2012).

Diversified production is also not possible without a policy and market enabling crop diversity and rewarding the contributions toward biodiversity and CCAM. Sustainable out-scaling of CCAM and to achieve large-scale landscape adoption depends on linkage to functional markets for inputs and outputs (Sommer et al. 2014). There is need for more attention via policy and extension on value chains, and more focused development resources from donors, governments, nongovernmental organizations (NGOs), and national and international research and development organizations to support participatory dissemination and upscaling of CCAM approaches (Friedrich et al. 2012). Investment policies should focus along value chains from technology at farm level, up to processing, and the availability of specifically forage legume seeds and tree seedlings.

Certification, that is, labeling of CCAM based agriculture, might be an option to support their implementation if linked with subsidies. In case of OA certification, schemes protecting the production status is currently only a half solution to increase the share of CCAM approaches in practice, due to the fact that only farmers with products for export are able to partly cover the relatively high certification costs. For local markets, certified products with a premium price are not adapted to the economic conditions of local population. Apart from specific investment costs, the successful implementation of CCAM methods would increase production and thus farmers income, making additional price premiums less important. An ecosystem services-based payment scheme might be an option; however, the administrative costs can be currently higher than the money reaching the farmers pocket.

To make an impact, integrated participatory catchment strategies are expected to be most efficient in terms of adoption of CCAM methods, via technology, advisory and financial support, and the implementation of an impact monitoring (Scherr et al. 2012).

Sensitivity must be given of the vulnerable and marginalized (women and poorer households, for instance) to avoid the exclusion and a social gradient (Feder et al. 2010). Multiple communication tools are discussed as a precondition for successful implementation of CCAM (Leeuwis and Aarts 2011). To bring together broad experience for learning and encouraging innovation, communication platforms should be established including farmers' organizations, advisory services, and universities. Social acceptance by referents, and shared learning processes, plays a key role in this regard (Lalani et al. 2017).

National policies have to frame this process, for example, to strengthen the establishment of a seed and seedling sector specifically for forage legumes and legume trees, while regional land locally sensitive policy instruments serve for finetuning of product specific value chains (Zougmoré et al. 2016).

Conclusions

This chapter summarizes the status and potential of smallholder farmers' application of CCAM practices and some observations on the institutional environment, with reference mainly to the subtropical/tropical environments of East African smallholder farming.

Lessons learned are that many CCAM relevant arable and plant cultivation methods are available, but most of these are still not in place. It is obvious that all CCAM methods contribute to an increase of productivity, with minor exceptions, where systems need further development.

With respect to agricultural production systems where the integration of CCAM methods is far advanced, conservation agriculture (CA) builds on the use of farm inputs including nitrogen fertilizer, soluble mineral fertilizers, herbicides, and chemical pesticides, while organic agriculture (OA) excludes these inputs. To what extent these inputs are a precondition for successful CA management is a matter of controversy. CCAM methods are practically similar in both approaches – CA and OA – while the degree of their implementation and favored methods might be different.

The amount of biomass production and nitrogen by nitrogen fixing crops is a key driver of CCAM and an increase of farm productivity. Both can serve as general indicators to assess farms, that is, catchments in their progress to cope with CC and to boost productivity, while it is less the share of mineral fertilizer. Most relevant crop groups with highest impact are forages legumes, forage grain legumes, and leguminous alley crops, but also biomass rich hybrid grasses, that profit indirectly from the forage legume nitrogen, for example, via slurry. These crop groups are the ones that increase soil fertility, reduce soil erosion and increase water holding capacity, biological pest and disease control, provide the feed biomass for animal husbandry systems, and guarantee an increase of cash crop yield and quality. Critical is the low supply of specifically forage legumes in seed markets.

There are still open questions, specifically concerning the GHG of mulching systems, pest and disease management and weed control, adapted technology for small scale farming and catchment strategies for a synergetic and efficient management. But those aspects are more part of a site and farm specific fine tuning than of fundamental lack of knowledge, which currently characterizes one of the main innovation barriers.

The dramatic negative trends of soil quality, biodiversity, and CC speak a clear language that the agricultural systems must bid farewell to one-sided systems. There is a need for a fundamental change, that is, resetting of the current farming practices, in other words, adaptation is no longer the appropriate term but the comprehensive transformation of the farming system as a whole. This is only possible if there is an enabling institutional and policy environment that supports agricultural research, advisory services, and education oriented to farmers' needs to adapt their farming system, making it more robust against drought and floods through the establishment of biomass and nitrogen producing cropping systems, and is reducing the farm specific negative impact on CC. This transformation process depends on assuring that best practices, adapted to local conditions, can demonstrate a reduction of labor, and a serious increase of farm income. It is recommended to apply a more systemic view in developing farming systems and related value chains, instead of focusing on single methods, that is, products. Transformation towards biodiverse crop rotations and biomass management also asks for the respective markets, consumer demands, and processing units, for example, for an increased milk production as a result of increased forage production, a dairy structure, and adaptions, that is, transformations of human diets. These final remarks clearly inform about the need for systemic transformation of food systems for successful CCAM management.

References

Achieng J, Ouma G, Odhiambo G, Muyekho F (2010) Effect of farmyard manure and inorganic fertilizers on maize production on Alfisols and Ultisols in Kakamega, western Kenya. Agric Biol J N Am 1(4):430–439

Adamtey N, Musyoka MW, Zundel C, Cobo JG, Karanja E, Fiaboe KK, ... Berset E (2016) Productivity, profitability and partial nutrient balance in maize-based conventional and organic farming systems in Kenya. Agric Ecosyst Environ 235:61–79

Awazi NP, Tchamba MN (2019) Enhancing agricultural sustainability and productivity under changing climate conditions through improved agroforestry practices in smallholder farming systems in Sub-Saharan Africa. Afr J Agric Res 14(7):379–388

Branca G, Lipper L, McCarthy N, Jolejole MC (2013) Food security, climate change, and sustainable land management. A review. *Agron Sustain Dev* 33(4):635–650

Broom D, Galindo F, Murgueitio E (2013) Sustainable, efficient livestock production with high biodiversity and good welfare for animals. Proc R Soc B Biol Sci 280(1771):20132025

Bunch R (2017) How can we cover millions of hectares with conservation agriculture in Africa. Conservation agriculture for Africa: building resilient farming systems in a changing climate. CABI: Boston, 139

Bunderson WT, Jere ZD, Thierfelder C, Gama M, Mwale BM, Ng'oma SW, ... Mkandawire O (2015) Implementing the principles of conservation agriculture in Malawi: crop yields and factors affecting adoption. Conservation Agriculture for Africa: Building Resilient Farming Systems in a Changing Climate

Cordingley JE, Snyder KA, Rosendahl J, Kizito F, Bossio D (2015) Thinking outside the plot: addressing low adoption of sustainable land management in sub-Saharan Africa. Curr Opin Environ Sustain 15:35–40

Dale D (2010) Sustainable land management technologies and approaches in Ethiopia. SLMP. Natural Resources Management Sector, MOARD, Addis Ababa

Dupar M (2019) Intergovernmental Panel on Climate Change (2019). Climate Change and Land: An IPCC Special Report on climate change, desertification, land degradation, sustainable land management, food security, and greenhouse gas fluxes in terrestrial ecosystems. Rome

Elwell H, Stocking M (1988) Loss of soil nutrients by sheet erosion is a major hidden farming cost. Zimb Sci News 22(7):8

Erenstein O, Sayre K, Wall P, Hellin J, Dixon J (2012) Conservation agriculture in maize-and wheat-based systems in the (sub) tropics: lessons from adaptation initiatives in South Asia, Mexico, and Southern Africa. J Sustain Agric 36(2):180–206

FAO (2011) Save and grow - a policymaker's guide to the sustainable intensification of smallholder crop production. FAO, Rome

FAO (2012) The State of Food and Agriculture. FAO, Rome

Feder G, Anderson JR, Birner R, Deininger K (2010) Promises and realities of community-based agricultural extension. In Community, Market and State in Development. Springer, New York (pp. 187–208)

Forster D, Adamtey N, Messmer MM, Pfiffner L, Baker B, Huber B, Niggli U (2012) Organic agriculture—driving innovations in crop research. In Agricultural sustainability-progress and prospects in crop research. Elsevier, pp 21–46

Franzel S, Carsan S, Lukuyu B, Sinja J, Wambugu C (2014) Fodder trees for improving livestock productivity and smallholder livelihoods in Africa. Curr Opin Environ Sustain 6:98–103

Freyer B (2019) The role of the crop rotation in organic farming. In: Köpke U (ed) Improving organic crop cultivation. Burleigh Dodds Science Publishing Limited, Cambridge, MA, pp 547–568

Friedrich T, Derpsch R, Kassam A (2012) Overview of the global spread of conservation agriculture. Field actions science reports. J Field Actions (Special Issue 6), p. 1–7

Garrity DP (2017) How to make conservation agriculture evergreen. In Kassam AH, Mkomwa S, Friedrich T (eds) Conservation agriculture for Africa. Building resilient farming systems in a changing climate. CABI: Boston. pp 167–182

Ghimire SR, Njarui DM, Mutimura M, Cardoso Arango JA, Johnson L, Gichangi E, ... Rao IM (2015) Climate-smart Brachiaria for improving livestock production in East Africa: Emerging opportunities. Paper presented at the Sustainable use for grassland resources for forage production, Range Management Society of India, Jhansi, India

Giller KE, Corbeels M, Nyamangara J, Triomphe B, Affholder F, Scopel E, Tittonell P (2011) A research agenda to explore the role of conservation agriculture in African smallholder farming systems. Field Crop Res 124(3):468–472

Gómez-Carabalí A, Idupulapati Madhusudana R, Ricaute J (2010) Differences in root distribution, nutrient acquisition and nutrient utilization by tropical forage species grown in degraded hillside soil conditions[1]. Acta Agronómica 59(2):197–210

Gonzalez-Sanchez EJ, Veroz-Gonzalez O, Conway G, Moreno-Garcia M, Kassam A, Mkomwa S, ... Carbonell-Bojollo R (2019) Meta-analysis on carbon sequestration through conservation agriculture in Africa. Soil Tillage Res 190:22–30

Grant MJ, Booth A (2009) A typology of reviews: an analysis of 14 review types and associated methodologies. Health Inform Libr J 26(2):91–108

Gurtner M, Liniger H, Studer R, Hauert C (2011) Sustainable land management in practice: Guidelines and best practices for Sub-Saharan Africa. FAO, Rome

Hillocks R (1998) The potential benefits of weeds with reference to small holder agriculture in Africa. Integr Pest Manag Rev 3(3):155–167

Hobbs PR, Govaerts B (2010) How conservation agriculture can contribute to buffering climate change. Climate change and crop production. CABI, Cambridge, p. 177–199

Holden ST (2018) Fertilizer and sustainable intensification in Sub-Saharan Africa. Glob Food Sec 18:20–26

IPCC (2018) Special Report Climate Change and Land. IPCC, Rome

Kassam A, Basch G, Friedrich T, Gonzalez E, Trivino P, Mkomwa S (2017) Mobilizing greater crop and land potentials sustainably. Hung Geograph Bull 66(1):3–11

Khan Z, Pickett JA, Wadhams LJ, Hassanali A, Midega CA (2006) Combined control of Striga hermonthica and stemborers by maize–Desmodium spp. intercrops. Crop Prot 25(9):989–995

Khan Z, Amudavi DM, Midega CA, Wanyama JM, Pickett JA (2008a) Farmers' perceptions of a 'push–pull'technology for control of cereal stemborers and Striga weed in western Kenya. Crop Prot 27(6):976–987

Khan Z, Midega CA, Amudavi DM, Hassanali A, Pickett JA (2008b) On-farm evaluation of the 'push–pull'technology for the control of stemborers and striga weed on maize in western Kenya. Field Crop Res 106(3):224–233

Khan Z, Midega CA, Pittchar JO, Murage AW, Birkett MA, Bruce TJ, Pickett JA (2014) Achieving food security for one million sub-Saharan African poor through push–pull innovation by 2020. Philos Trans R Soc B: Biol Sci 369(1639):20120284

Khan Z, Midega CA, Hooper A, Pickett J (2016) Push-pull: chemical ecology-based integrated pest management technology. J Chem Ecol 42(7):689–697

Khan Z, Midega CA, Pittachar J, Murage A, Pickett J (2017) Climate-smart push-pull: a conservation agriculture technology for food security and environmental sustainability in Africa. In: Conservation agriculture for Africa: building resilient farming systems in a changing climate. BABI, Wallingford, pp 151–166

Koepke U (2003) Conservation agriculture with and without use of agrochemicals. Paper presented at the proceedings of the 2nd world congress on conservation agriculture, Iguassu Falls, Paraná, Brazil

Kumar S, Meena RS, Lal R, Yadav GS, Mitran T, Meena BL, … Ayman E-S (2018) Role of legumes in soil carbon sequestration. In Legumes for Soil Health and Sustainable Management. Springer. New York (pp. 109–138)

Lalani B, Dorward P, Kassam AH, Dambiro J (2017) Innovation systems and farmer perceptions regarding conservation agriculture in Cabo Delgado, Mozambique. AH Kassam, S. Mkomwa, & T. Friedrich, Conservation Agriculture for Africa: Building resilient farming systems in a changing climate, 100–126. CABI, Boston

Lasco RD, Delfino RJP, Catacutan DC, Simelton ES, Wilson DM (2014) Climate risk adaptation by smallholder farmers: the roles of trees and agroforestry. Curr Opin Environ Sustain 6:83–88

Leeuwis C, Aarts N (2011) Rethinking communication in innovation processes: creating space for change in complex systems. J Agric Educ Extension 17(1):21–36

Lotter D (2015) Facing food insecurity in Africa: why, after 30 years of work in organic agriculture, I am promoting the use of synthetic fertilizers and herbicides in small-scale staple crop production. Agric Hum Values 32(1):111–118

Mahon N, Crute I, Simmons E, Islam MM (2017) Sustainable intensification–"oxymoron" or "third-way"? A systematic review. Ecol Indic 74:73–97

Majumder K (2015) A young researcher's guide to a systematic review. Editage, Japan

Mason-D'Croz D, Sulser TB, Wiebe K, Rosegrant MW, Lowder SK, Nin-Pratt A, … Cenacchi N (2019) Agricultural investments and hunger in Africa modeling potential contributions to SDG2–Zero Hunger. World Dev 116:38–53

Mbow C, Smith P, Skole D, Duguma L, Bustamante M (2014) Achieving mitigation and adaptation to climate change through sustainable agroforestry practices in Africa. Curr Opin Environ Sustain 6:8–14

Mchunu CN, Lorentz S, Jewitt G, Manson A, Chaplot V (2011) No-till impact on soil and soil organic carbon erosion under crop residue scarcity in Africa. Soil Sci Soc Am J 75(4):1503–1512

Mdee A, Wostry A, Coulson A, Maro J (2019) A pathway to inclusive sustainable intensification in agriculture? Assessing evidence on the application of agroecology in Tanzania. Agroecol Sustain Food Syst 43(2):201–227

Meyer R (2010) Low-input intensification in agriculture chances for small-scale farmers in developing countries. Gaia-Ecol Perspect Sci Soc 19(4):263–268

Midega CA, Salifu D, Bruce TJ, Pittchar J, Pickett JA, Khan ZR (2014) Cumulative effects and economic benefits of intercropping maize with food legumes on Striga hermonthica infestation. Field Crop Res 155:144–152

Midega CA, Bruce TJ, Pickett JA, Pittchar JO, Murage A, Khan ZR (2015) Climate-adapted companion cropping increases agricultural productivity in East Africa. Field Crop Res 180:118–125

Mkonda MY, He X (2017) Conservation agriculture in Tanzania. In: Sustainable agriculture reviews. Springer, Berlin, pp 309–324

Monegat C (1991) Plantas de cobertura del suelo: características y manejo en pequeñas propiedades. Universidad Nacional Agraria, (UNA), Nicaragua Centro Nacional de Investigación y Documentación Agropecuaria, (CENIDA)

Müller A, Davis JS (2009) Reducing global warming: the potential of organic agriculture. FiBL, Frick, Switzerland

Murthy IK, Gupta M, Tomar S, Munsi M, Tiwari R, Hegde G, Ravindranath N (2013) Carbon sequestration potential of agroforestry systems in India. J Earth Sci Clim Chang 4(1):1–7

Murthy IK, Dutta S, Vinisha V (2016) Impact of agroforestry sytems on ecological and socioeconomic systems: a review. Glob J Sci Front Res: H Environ Earth Sci 16(5):15–27

Musyoka MW, Adamtey N, Muriuki AW, Cadisch G (2017) Effect of organic and conventional farming systems on nitrogen use efficiency of potato, maize and vegetables in the Central highlands of Kenya. Eur J Agron 86:24–36

Nalubwama SM, Mugisha A, Vaarst M (2011) Organic livestock production in Uganda: potentials, challenges and prospects. Trop Anim Health Prod 43(4):749–757

Nandwa S, Obanyi S, Mafongoya P (2011) Agro-ecological distribution of legumes in farming systems and identification of biophysical niches for legumes growth. In: Fighting poverty in Sub-Saharan Africa: the multiple roles of legumes in integrated soil fertility management. Springer, Dordrecht, pp 1–26

Ndambi OA, Pelster DE, Owino JO, De Buisonje F, Vellinga T (2019) Manure management practices and policies in sub-Saharan Africa: implications on manure quality as a fertilizer. Front Sustain Food Syst 3:29

Ngwira AR, Thierfelder C, Lambert DM (2013) Conservation agriculture systems for Malawian smallholder farmers: long-term effects on crop productivity, profitability and soil quality. Renew Agric Food Syst 28(4):350–363

Nyasimi M, Amwata D, Hove L, Kinyangi J, Wamukoya G (2014) Evidence of impact: climate-smart agriculture in Africa

Ochieng SA, Waiswa DC (2019) Pastoral education: the missing link in Uganda education system. Educ Res Rev 14(7):240–253

Partey ST, Zougmoré RB, Ouédraogo M, Campbell BM (2018) Developing climate-smart agriculture to face climate variability in West Africa: challenges and lessons learnt. J Clean Prod 187:285–295

Porter JR, Challinor AJ, Henriksen CB, Howden SM, Martre P, Smith P (2019) Invited review: intergovernmental panel on climate change, agriculture, and food—a case of shifting cultivation and history. Glob Chang Biol 25(8):2518–2529

Powlson DS, Stirling CM, Jat ML, Gerard BG, Palm CA, Sanchez PA, Cassman KG (2014) Limited potential of no-till agriculture for climate change mitigation. Nat Clim Chang 4(8):678–683

Pratt J, Henry E, Mbeza H, Mlaka E, Satali L (2002) Malawi agroforestry extension project marketing & enterprise program, main report. Malawi Agroforestry 47:139

Pretty J, Bharucha ZP (2015) Integrated pest management for sustainable intensification of agriculture in Asia and Africa. Insects 6(1):152–182

Reicosky DC (2015) Conservation tillage is not conservation agriculture. J Soil Water Conserv 70 (5):103A–108A

Rusinamhodzi L, Corbeels M, Nyamangara J, Giller KE (2012) Maize–grain legume intercropping is an attractive option for ecological intensification that reduces climatic risk for smallholder farmers in Central Mozambique. Field Crop Res 136:12–22

Sarvade S, Upadhyay V, Agrawal S (2019) Quality fodder production through silvo-pastoral system: a review. In: Dev I, Ram A, Kumar N, Singh R, Kumar D, Uthappa AR, Handa AK, Chaturvedi OP (eds) Agroforestry for climate resilience and rural livelihood. Scientific Publishers, Jodhpur, pp 345–359

Scherr SJ, Shames S, Friedman R (2012) From climate-smart agriculture to climate-smart landscapes. Agric Food Security 1(1):12

Schroth G, Balle P, Peltier R (1995) Alley cropping groundnut with Gliricidia sepium in Cote d'Ivoire: effects on yields, microclimate and crop diseases. Agrofor Syst 29(2):147–163

Schultze-Kraft R, Rao IM, Peters M, Clements RJ, Bai C, Liu G (2018) Tropical forage legumes for environmental benefits: an overview. Trop Grasslands-Forrajes Tropicales 6(1):1–14

Scialabba NE-H, Müller-Lindenlauf M (2010) Organic agriculture and climate change. Renew Agric Food Syst 25(2):158–169

Scopel E, Triomphe B, Ribeiro MdS, Séguy L, Denardin JE, Kochann R (2004) Direct seeding mulch-based cropping systems (DMC) in Latin America. Paper presented at the new directions for a diverse planet: proceedings for the 4th international crop science congress, Brisbane

Senyolo MP, Long TB, Blok V, Omta O (2018) How the characteristics of innovations impact their adoption: an exploration of climate-smart agricultural innovations in South Africa. J Clean Prod 172:3825–3840

Sheaffer CC, Seguin P (2003) Forage legumes for sustainable cropping systems. J Crop Prod 8 (1–2):187–216

Sietz D, Van Dijk H (2015) Land-based adaptation to global change: what drives soil and water conservation in western Africa? Glob Environ Chang 33:131–141

Sims B, Kienzle J, Mkomwa S, Friedrich T, Kassam A (2017) Mechanization of smallholder conservation agriculture in Africa; contributing resilience to precarious systems. Conservation Agriculture for Africa. Building Resilient Farming Systems in a Changing Climate; Kassam, AH, Mkomwa, S., Friedrich, T., Eds, 183–213. CABI, Boston

Smith A, Snapp S, Chikowo R, Thorne P, Bekunda M, Glover J (2017) Measuring sustainable intensification in smallholder agroecosystems: a review. Glob Food Sec 12:127–138

Sommer R, Piggin C, Haddad A, Hajdibo A, Hayek P, Khalil Y (2012) Simulating the effects of zero tillage and crop residue retention on water relations and yield of wheat under rainfed semiarid Mediterranean conditions. Field Crop Res 132:40–52

Sommer R, Thierfelder C, Tittonell P, Hove L, Mureithi J, Mkomwa S (2014) Fertilizer use should not be a fourth principle to define conservation agriculture: response to the opinion paper of Vanlauwe et al. (2014)'A fourth principle is required to define conservation agriculture in sub-Saharan Africa: the appropriate use of fertilizer to enhance crop productivity'. Field Crops Res 169:145–148

Teenstra E, de Buisonjé F, Ndambi A, Pelster D (2015) Manure Management in the (Sub-) Tropics: training manual for extension workers. Wageningen UR (University & Research centre) Livestock Research, Livestock Research Report 919

Tesfa A, Mekuriaw S (2014) The effect of land degradation on farm size dynamics and crop-livestock farming system in ethiopia: A Review. Open Journal of Soil Science, 2014, p. 1–5

Thierfelder C, Wall P (2010) Rotation in conservation agriculture systems of Zambia: effects on soil quality and water relations. Exp Agric 46(3):309–325

Thierfelder C, Cheesman S, Rusinamhodzi L (2013) Benefits and challenges of crop rotations in maize-based conservation agriculture (CA) cropping systems of southern Africa. Int J Agric Sustain 11(2):108–124

Thierfelder C, Bunderson WT, Jere ZD, Mutenje M, Ngwira A (2016) Development of conservation agriculture (CA) systems in Malawi: lessons learned from 2005 to 2014. Exp Agric 52(4):579–604

Thomas RJ, Asakawa N (1993) Decomposition of leaf litter from tropical forage grasses and legumes. Soil Biol Biochem 25(10):1351–1361

Tittonell P, Giller KE (2013) When yield gaps are poverty traps: the paradigm of ecological intensification in African smallholder agriculture. Field Crop Res 143:76–90

Traill S, Bell LW, Dalgliesh NP, Wilson A, Ramony L-M, Guppy C (2018) Tropical forage legumes provide large nitrogen benefits to maize except when fodder is removed. Crop Pasture Sci 69 (2):183–193

Tully K, Sullivan C, Weil R, Sanchez P (2015) The state of soil degradation in Sub-Saharan Africa: baselines, trajectories, and solutions. Sustainability 7(6):6523–6552

Vanlauwe B, Aihou K, Tossah B, Diels J, Sanginga N, Merckx R (2005) Senna siamea trees recycle Ca from a Ca-rich subsoil and increase the topsoil pH in agroforestry systems in the West African derived savanna zone. Plant Soil 269(1–2):285–296

Vanlauwe B, Coyne D, Gockowski J, Hauser S, Huising J, Masso C, ... Van Asten P (2014a) Sustainable intensification and the African smallholder farmer. Curr Opin Environ Sustain 8:15–22

Vanlauwe B, Wendt J, Giller KE, Corbeels M, Gerard B, Nolte C (2014b) A fourth principle is required to define conservation agriculture in sub-Saharan Africa: the appropriate use of fertilizer to enhance crop productivity. Field Crop Res 155:10–13

Wall P (2017) Conservation Agriculture: Growing More with Less–the Future of Sustainable Intensification. Conservation agriculture for Africa, CABI, Boston. p. 30–40

Wezel A, Soboksa G, McClelland S, Delespesse F, Boissau A (2015) The blurred boundaries of ecological, sustainable, and agroecological intensification: a review. Agron Sustain Dev 35 (4):1283–1295

Wilkins RJ (2008) Eco-efficient approaches to land management: a case for increased integration of crop and animal production systems. Philos Trans R Soc B: Biol Sci 363(1491):517–525

Willer H, Lernoud J (2019) The world of organic agriculture. Statistics and emerging trends 2019: Research Institute of Organic Agriculture FiBL and IFOAM Organics International

Wilson G, Kang B, Mulongoy K (1986) Alley cropping: trees as sources of green-manure and mulch in the tropics. Biol Agric Horticulture 3(2–3):251–267

Wynants M, Kelly C, Mtei K, Munishi L, Patrick A, Rabinovich A, ... Boeckx P (2019) Drivers of increased soil erosion in East Africa's agropastoral systems: changing interactions between the social, economic and natural domains. Regional Environmental Change, 19, p. 1909–1921

Xie H, Huang Y, Chen Q, Zhang Y, Wu Q (2019) Prospects for agricultural sustainable intensification: a review of research. Land 8(11):157

Zougmoré R, Partey S, Ouédraogo M, Omitoyin B, Thomas T, Ayantunde A, ... Jalloh A (2016) Toward climate-smart agriculture in West Africa: a review of climate change impacts, adaptation strategies and policy developments for the livestock, fishery and crop production sectors. Agric Food Security 5(1):26

Zougmoré RB, Partey ST, Ouédraogo M, Torquebiau E, Campbell BM (2018) Facing climate variability in sub-Saharan Africa: analysis of climate-smart agriculture opportunities to manage climate-related risks. Cahiers Agricultures (TSI) 27(3):1–9

4

Africa–European Union Climate Change Partnership

Oluwole Olutola

Contents

Abstract

The need to heighten climate action momentum is a key outcome of the Climate Action Summit organized by the United Nations (UN) in September, 2019. The same concern reverberated in most of the presentations and discussions at the twenty-fifth Conference of Parties (COP 25) – the annual climate summit under the United Nations Framework Convention on Climate Change (UNFCCC). This chapter seeks to investigate the relevance of the call for more climate action in terms of what further climate priorities and strategies are required in the context of the existing climate change partnership between Africa and the European Union (EU). It relies on liberal institutionalism as its theoretical framework and data from a range of purposely selected secondary sources as reference points. Beyond arguing the case for more climate action to further strengthening the Joint Africa-EU Strategy (JAES), particularly in the area of environmental partnership, this chapter emphasizes the need to align the required further climate action with the mitigation goals of the Paris Agreement and the UN transformative initiatives on

O. Olutola (✉)
University of Johannesburg, Johannesburg, South Africa

the global climate action. It concludes with an insight into some policy recommendations, including the need for a dedicated and regional-based approach in tackling Africa's climate change beyond the conventional worldwide UNFCCC (United Nations Convention on Climate Change) framework that has failed to deliver tangible results for some time past.

Keywords

Climate change · Liberal institutionalism · COP25 · Joint Africa–EU Strategy · Paris Agreement

Introduction

Addressing climate change as an existential threat to this generation (UNCC 2019a) and the future generation given its transgenerational implications is more urgent than ever. In recognizing this growing climate concern, the United Nations (UN) convened a global climate action held in New York on 23 September 2019. The primary objective was aimed at mobilizing wide-range support for the multilateral climate change process. In the end, the summit emphasized the need to increase mitigation ambition as well as accelerate climate action involving a range of stakeholders – state and nonstate alike, including multilateral entities (UN 2019). Less than 3 months after, similar concern reverberated in most of the speeches and statements given at the twenty-fifth Conference of Parties (COP 25) – the annual climate summit under the United Nations Framework Convention on Climate Change (UNFCCC) – which took place in Madrid from 2 to 13 December 2019. Generally described as the launchpad for significantly more climate ambition, COP 25 also ended with a call for improved climate action (CarbonBrief 2019).

This separate but joint call for more climate ambition and action could not have happened at a better time, considering the mounting threats of climate change to the global system, regional and subregional entities. Besides, the call is consistent with the mitigation goals of the Paris Agreement that was decided at COP 21 in December 2015 as the first-ever universally agreed climate deal after more than two decades of unduly prolonged negotiations that characterized the previous COP meetings (Amusan and Olutola 2016). Under the Paris Agreement, state entities commit to ensure that the global average temperature is pegged to 2 °C above preindustrial levels and, if possible, further down to 1.5 °C still above preindustrial levels (UNFCCC 2015). However, the collective efforts to meet the set mitigation target are currently insufficient (Boyd et al. 2015; Schleussner et al. 2016). Recent finding shows that there must be a cut in carbon emissions to about 45% and net zero by 2030 and 2050, respectively, to save this century from the irreversible and catastrophic impacts of climate change (IPCC 2018).

From the outcomes of the aforementioned summits and the IPCC carbon cut projection, it is deduced that both the past and current efforts – at all levels – to combat climate change remain inadequate and far less than what should be the case. The significant attention drawn to the mounting dangers of climate change and the multistakeholder approach in dealing with the phenomenon as a common enemy and global emergency in particular points to establish the summits' acknowledgment of multilateral entities as important rallying points for the desired enhanced climate action.

This chapter presents the case of Africa and the European Union (EU) partnership focusing on climate change as one of the priorities in the relationship between the two continental partners. It seeks to examine the relevance or otherwise of the call for more climate ambition and action in this particular case, and what further climate priorities and strategies, if any, are required. The chapter is structured into five sections as follows: section one contains the above introduction; section two systematically examines the key constructs of liberal institutionalism as a theoretical basis for this study; section three gives an overview of the UN climate action summit and COP 25 in the context of the Paris Agreement; section four appraises Africa–EU climate change partnership in light of the call for more climate action; and section five closes with a conclusion including an insight into key policy recommendations.

Liberal Institutionalism

Liberal institutionalism represents one of the theoretical strands of the liberal school of thought. Generally, liberalism introduced new paradigm of debates to the body of international relations theories, as it underscores the relevance of nonmilitary (security) approach to handling issues and matters of common priority within the international system. Liberalists' main concern is to construct a model of international relations with capacity to mitigate the unchecked use of military force as a foreign policy instrument by state actors.

For most liberal institutionalists, cooperation between state and nonstate actors remains the most important and mutually beneficial ordering feature of the international system (Keohane 1984; Keohane and Martin 1995). This interstate cooperation is facilitated through international institutions and regimes, defined as a set of implicit or explicit principles, norms, rules, and decision-making procedures around which actors' expectations converge regarding any aspect of international relations (Krasner 1983: 2). The implication is that international institutions and regimes are conceived as the primary means of limiting the power of states at both domestic and international levels, thereby mitigating anarchy in the international system (Burchill 2005: 65).

Besides, liberal institutionalism places emphasis on international institutions which have the ability to help overcome selfish state behavior by bringing them together in a cooperative manner in pursuit of shared foreign policy objectives otherwise unattainable in isolation. In other words, international institutions serve as entities for mobilization networks, within which transgovernmental policy coordination and coalition building could take place (Keohane and Nye 1987: 738). In addition to providing multilateral platforms through which states deal with collective action problems that threaten stable patterns of cooperation, international institutions also perform such roles as coordination and monitoring which together make them to become "valuable foundation" for international cooperation (Martin 2007: 111).

Another key assumption of liberal institutionalism is the existence of multiple channels of contact through which states and societies are interconnected (Keohane and Nye 1987: 731). This brings to focus the term complex interdependence and the argument that the ranking of global issues as high and low politics is uncalled for, particularly in a world of multiple issues imperfectly linked and characterized by transnational and transgovernmental coalitions (Grieco 1988: 490).

However, liberal institutionalists agree that interstate cooperation is constrained by cheating and noncompliance with international agreements given the self-enforcing and anarchic nature of the international system. The situation is further worsened by the lack of guarantee to ensure that state individual tendencies to maximize the gains of cooperation at the expense of other participating actors are regulated in such a way that benefits are shared equally.

These shortcomings notwithstanding, liberal institutionalists strongly believe that cooperation between states is still possible even though it is something that happens gradually. In their analysis, cooperation would first be achieved in technical areas where it was mutually convenient and, if successful, could be extended to other functional areas of mutual benefits (Burchill 2005: 64). Indeed, the emphasis on international institutions as an important rallying point for interstate cooperation and, also, a potentially effective mechanism for containing global emergencies brings to relevance the focus on Africa and the EU.

While the Africa–EU climate change partnership is not a standalone multilateral institution in itself, it represents a key element and one of the important thematic priorities under the Joint Africa–European Union Strategy (JAES) adopted at the second EU–Africa summit in Lisbon in 2007. Section four of this chapter provides more explanations on the JAES. However, it is important to stress that Africa and Europe as two key multilateral partners depend on the instrumentality of the African Union and the EU to provide the needed institutional framework for the implementation of JAES and, more specifically, the partnership on climate change and other related climate activities being discussed in this chapter.

Paris Agreement, UN Climate Action Summit, and COP 25

As a special creation of the UN, the UNFCCC is responsible for the global negotiations in response to climate change. Since its establishment in 1992, the global climate change process under the UNFCCC framework has experienced a back-and-forth approach to climate negotiations and, so, action. But after nearly two and half decades of interrupted negotiations, the Paris Agreement was agreed as a globally accepted climate action plan in 2015. By the agreement, state entities commit to ensure that the global average temperature is pegged to 2 °C above preindustrial levels and, if possible, to 1.5 °C above preindustrial levels (UNFCCC 2015).

To attain this long-term mitigation ambition, each party to the UNFCCC is under an obligation to develop and commit to a nationally determined contribution (NDC), which should be communicated to the UNFCCC secretariat and progressively maintained. The provisions of the agreement include a ratchet-up mechanism to periodically review and update the NDCs every 5 years effectively from 2018 upward. While the unanimity exists in terms of perception of NDCs as a collective response towards achieving the Paris climate goals, the common understanding of what exactly constitute the NDCs is still lacking among the UNFCCC parties. Besides, efforts to ensure its transparency, particularly in the context of national climate framework and objectives, are still quite challenging.

Besides, the current emission reduction pledges captured in the NDCs for the period up until 2030 though represent progress compared with "business as usual," but insufficient to secure the achievement of the mitigation goal set by the Paris Agreement (Boyd et al. 2015; Schleussner et al. 2016; UN 2019). Nevertheless, it forms the basis of any serious global struggle against climate change, especially on the part of state actors. It also points to establish that collective efforts beyond the UNFCCC are, indeed, needed to effectively address climate change. Stabilizing the global climate at safe levels requires wider international cooperation to complement the global climate change process within the UNFCCC (Moncel and van Asselt 2012).

Unfortunately, not so much of the mitigation ambition proposed under the Paris Agreement has been achieved. The full implementation of the Paris Agreement is yet to be actualized, as efforts are still ongoing at the level of the annual Conference of Parties to finalize its operational guidelines. The latest in the series was COP25 held in 2019 as the Launchpad for significantly more climate ambition. It is important to note that the Trump-led US in mid-2017 formally disclosed the country's intention to withdraw from further participating in the Paris Agreement (Lawrence and Wong 2017). While there is no consensus in research yet as to whether the US withdrawal represents an opportunity or a setback for the Paris Agreement in particular and the global climate action in general, some scholars have raised concerns around the potential damage that could result from the US nonparticipation in raising finance to support global climate action (Olutola 2020; Urpelainen and van de Graaf 2018).

Yet, the worsening impacts of climate variation are becoming increasingly evident in some parts of the world. Recent cases include the Hurricane Dorian that struck the Bahamas and Cyclone Idai landfall in Mozambique with their attendant unprecedented catastrophes. Obviously worried by this growing severity of climate change impacts, the UN as a universal body gathered together wide-ranging stakeholders – state and nonstate – in what was dubbed the global climate action summit held in December 2019. The intention was to provide support for the Paris Agreement and the UN 2030 Agenda for sustainable development. In particular, it offered state participants a unique opportunity to discuss how best to enhance their respective NDCs by 2020 (short-term) in keeping faith with 2030 (mid-term) mitigation goal of 45% emission reduction as well as the 2050 (long-term) mitigation objective of net zero relative to GHG emissions.

The summit focused on nine key areas where urgent climate action is required. These include energy transition; climate finance and carbon pricing; resilience and adaptation; nature-based solutions; mitigation strategy; among others (UN 2019: 3). Buried under 12 themes, the summit's climate action objectives are expected to be achieved through transformative initiatives for which stakeholders would be held responsible. These transformative initiatives include: the need for improved climate finance as a key element for the transition to net-zero emissions and climate resilient economies; pledges to decarbonize investment portfolios and systematically include environmental impacts in investment decision-making; setting limits for the use of coal, or phase it out altogether, including the development of a collective support system to help provide developing countries with the option of exiting coal; plans to eliminate deforestation, preserving biodiversity, and restoration of natural ecosystems particularly through planting of trees; integration of climate risks and resilience initiatives in decision-making systems and national development frameworks across the continent, including climate resilient development pathways for least developed countries (LCDs); and provision of insurance for the most vulnerable and support to prevent climate-related disasters, among others. More importantly, the summit succeeded in establishing an all-encompassing steering committee to provide strategic guidance and oversight of its planned action and activities as well as two advisory groups – science and ambition – to provide technical expertise.

Aside from stressing the need for the urgency of climate action in the identified areas, the summit called attention to the strategic importance of renewed leadership at all levels and across the board including collaboration between relevant stakeholders. Its multistakeholder transformative initiatives with commitments from 70 and 75 countries – mostly small and developing countries responsible for far below 15% of aggregate carbon emissions worldwide – to work towards more aggressive NDCs and net-zero emissions by 2020 and 2050 respectively, are nevertheless remarkable. Granted that the agreed initiatives are no doubt consistent with the Paris mitigation objectives, it is of concern that the summit could not secure concrete and immediate mitigation pledges from the world's leading GHG emitters – mostly the G20 countries (including the full EU) which together produce close to 30 kilotons of CO_2 annually, as of 2015, thereby responsible for about 81% of all global carbon emissions (Globalist 2018).

Unfortunately, the COP25 – which was to provide a critical platform for the operationalization of the Paris Agreement with the year 2020 set as the deadline – fell short of expectations. Despite the momentum ignited by the UN climate action summit, the once in a year climate meeting could not achieve much. The climate ambition aglliance (UNCC 2019b) presented during the meeting is chiefly a recapitulation of the multistakeholder pledges made at the UN climate action summit. No consensus was reached regarding the planned increase in mitigation commitments, while virtually all other outstanding issues emanating from the Paris Agreement were also left unresolved. These issues range from failure to secure: increased NDCs pledges, especially from the world's biggest emitters; the final decision on the rulebook, regarded as the operating manual for the implementation of the Paris climate deal; specific operating guidelines for loss and damage; and new and enhanced climate finance goals, among others.

This lack of progress is worrisome and, certainly, not a good complementarity of the UN transformative initiatives concerning the global climate action. Besides, it exacerbates the concern raised in the emissions gap report that the existing NDCs, even if met, would not be enough to deliver the Paris mitigation goal. Based on the report, emissions need to reduce to 2.7% each year from 2020 to 2030 and 7.6% each year on the average for the 2 °C and the 1.5 °C goals, respectively, to meet the Paris Agreement's mitigation target (UNEP 2019: 26). Hence, the urge for increased ambition pledges is critical to closing the gap between the emission targets captured in the NDCs currently and the mitigation goal set by the Paris Agreement.

Just as COP25 was winding up, the EU signaled its resolve to achieve net-zero emissions by 2050. Encapsulated in what is known as the European Green Deal, the EU seeks to commit about 25% of its long-term budget to climate-related objectives. As a deliberate strategy to boost the EU's NDC pledge for 2030, the deal contains a proposal to reduce the bloc's carbon emissions from its current target of 40% to a higher target of at least 50% and towards 55% compared with 1990 levels (Claeys et al. 2019; EC 2019). If this deal is actualized, coupled with the fact that other key emitters – most especially the USA – are showing no indication to seriously commit to any increased mitigation plan, the EU would have once again reestablished its pivotal role in providing leadership to the global action against climate change. This self-assumed responsibility brings to focus the climate change partnership between the EU and Africa (another continent that is at the center of any discussions on the global climate action because of its high exposure to climate change impacts and little or no capacity in terms of adaptation).

Africa–EU Climate Change Partnership: A Revisit

The partnership between Africa and the EU was launched in 2000 – two decades ago – at the maiden edition of the Africa–EU meeting in Cairo. Seven years after, the two partners at the second edition in Lisbon adopted a Joint Africa–EU Strategy (JAES). The JAES represents the guiding instrument for the overarching long-term and political framework of the collaboration between the two continental entities (EU 2007). It outlines the basic principles (ownership, partnership, and solidarity) and general objectives of the partnership. These include a resolution on the part of the two partners to formalize the strategic partnership by moving away from the usual donor and recipient – give and take – approach; treat Africa as one entity; enhance their partnership at all levels on the basis of jointly identified mutual and complementary interests; and take their multilateral engagement to a new strategic level with reinforced policy dialogues and action plans, among other objectives (Bach 2010; EU 2013–2019).

Interestingly, climate change (and the environment) made the list as one of the thematic priorities of common concern in the Africa–EU partnership. Others include peace and insecurity; democratic governance and human rights; regional economic integration, trade, and infrastructure; millennium development goals; energy; mitigation, mobility, and employment; and science, information society, and space. Africa–EU climate change partnership can be viewed from at least two perspectives: the collaborative efforts of the two partners towards the global climate change process within the UNFCCC and the willingness on the part of the two partners to work together to combat climate change as a common enemy. Even though the vision of a joint agenda/position on climate change could not be achieved as envisaged partly due to lack of clarity on their common interests and internal divisions, the JAES's climate change partnership perhaps succeeded in building a common understanding of various climate-related issues and of their respective positions in the UNFCCC multilateral negotiations (Tondel et al. 2015).

The partnership has produced some level of significant progress over the years, as manifested in the launch of several climate-linked initiatives and programs. These include TerrAfrica, the Great Green Wall for the Sahara and the Sahel Initiative (GGWSSI), and Climate for Development in Africa (ClimDev-Africa). While TerrAfrica was created in 2005 as a platform for better coordination of efforts geared towards the upscaling of finance and mainstreaming of effective and efficient country-driven sustainable land and water management (SLWM) across the continent (NEPAD 2019), the GGWSSI was launched in 2007 as a "bulwark against the encroaching desert" (Bilski 2018), thereby strengthening climate resilience in Africa. ClimDev-Africa was designed as a tripartite arrangement of the United Nations Economic Commission for Africa (UNECA), the African Union Commission (AUC), and the African Development Bank (AfDB) through ClimDev Special Fund in 2010. Essentially, it aims to provide a solid foundation for an appropriate regional climate change response (UNECA 2014).

In particular, TerrAfrica and GGWSSI contributed significantly to strengthening the collaboration between the two partners, especially in the areas of sustainable land management and fight against desert encroachment in sub-Saharan Africa, respectively. Similarly, the EU financial intervention of €8 million through the ClimDev-Africa initiative was instrumental to the establishment of the African Climate Policy Center (ACPC) in Addis Ababa in 2012 and, by extension, the development of climate-based knowledge in support of policy-making in Africa (EU 2014: 24–25). By 2013, a €28 million contribution to ClimDev-Africa was launched to provide support (financial and technical) to the African Union (AU) – as the continent's collective representative – and many of its member states to enhance their capacities to make climate-sensitive policies. In 2015, the EU introduced another funding package amounting to €80 million to build disaster resilience in sub-Saharan Africa (EC 2015). Since the rebirth of the Organization of African Unity (OAU) as AU in 2002, Africa has received climate-related EU aid amounting to €3.7 billion (Khadiagala 2018: 440).

Many poor African countries with relatively high vulnerability to climate change (Chad, Cape Verde, Democratic Republic of Congo, Djibouti, Mali, Mauritania, Mozambique, Sudan, and Uganda, among others) have particularly benefited one way or the other from the EU global ecological charity administered through the Global Climate Change Alliance (GCCA 2018). The GCCA was created in 2007 to support climate change projects and programs in the world's most climate-vulnerable countries of which many are found in Africa. It provides technical and financial support for national, multicountry, and regional climate change projects and programs using a set of eligibility criteria (Miola et al., 2015). The GCCA is focused on five priorities, namely: mainstreaming climate change into poverty reduction and development strategies; adaptation, building on the National Adaptation Programs of Action (NAPAs) and other national plans; disaster risk reduction (DRR); reducing emissions from deforestation and forest degradation (REDD); and enhancing participation in the Global Carbon Market and Clean Development Mechanism (CDM).

Despite the progress recorded, climate change remains a challenging thematic priority for the EU–Africa partnership. For the most part, the partnership has been plagued by some issues ranging from cumbersome institutional structure; inefficient policy processes; mistrust; capacity differentials; lack of clarity on shared purpose and priorities; a deficit of political support on both sides; and the Brexit anxiety, among others. To address these issues and achieve common objectives in the important area of climate change, Africa and the EU must work more closely and commit to more climate action in line with the UN transformative initiatives.

The call for more climate action is therefore a wake-up opportunity for Africa and the EU to take deliberate action towards deepening the existing climate change collaboration between the two partners. Africa and the EU share affinities in the important aspects of their historical connections, geographical closeness, political vision, and interests, including the great potential for a common future. Up until this period, the two partners have played a pivotal role in the global fight against climate change. It is high time to consolidate their joint efforts and continue to take the lead in the global climate change struggle. One way to achieve this is by bringing the current approach in terms of climate action and strategies within the Africa–EU partnership into alignment with the agreed transformative initiatives on the global climate action. More specifically, it is high time that the Africa–EU climate change partnership complements the efforts of the UN steering committee on the global climate action, especially in providing strategic guidance and oversight of the implementation of the global transformative initiatives as they affect Africa.

In addition, Africa and the EU need to commit to a common climate change agenda and joint implementation framework that not only support the components of the UN transformative initiatives, but also consistent with the mitigation goal set by the Paris Agreement. Achieving this may face with the challenge of difference in priorities. As a marginal contributor to the global carbon emissions and, ironically, a core victim of climate change adverse impacts, Africa over the years has been consistent in its advocacy of adaptation bailout. The AU as the continent's collective representative minced no words in stating this regional climate change position in its Agenda 2063 (AUC 2015). While the EU has no doubt demonstrated support for the continent's adaptation priority, its primary focus like other developed parties to the UNFCCC is geared towards addressing mitigation in the form of emissions' reduction. This priority gap needs to be addressed.

Narratives about Africa and Europe are changing in recent years. The African continent, for instance, have demonstrated remarkable progress in some aspects such as governance and democratic accountability, human development, and sustained domestic economic growth. A case in point regarding changes in Europe is no doubt the Brexit phenomenon. There is therefore the need to adjust the EU–Africa relations in the context of these new developments. Africa–EU climate change policies in particular should be driven by common interests and objectives, with clearly defined priorities and action plans that recognize differences regarding the strengths and weaknesses of individual partners. More desirable is a balanced Africa–EU climate change relation. While it is important that financial and technical supports should be provided to African countries to enable them fulfill their climate action pledges, Africa cannot continue to depend entirely on bailout in its efforts to cope and adapt to climate change under the excuse of extreme vulnerability. A truly multilateral partnership entails collective action and shared responsibility in all aspects. With no prejudice to the fact that many African countries are relatively poor and faced with daunting challenges of sustainable development, it is time for Africa to stop paying lip service to the mantra "African solutions to African problems."

Actions in terms of climate action and strategies within the Africa–EU partnership should not only be aligned with the transformation initiatives, but also and above all, be structured around a dedicated and regional approach. This should go beyond the conventional worldwide UNFCCC (United Nations Convention on Climate Change) framework that has failed to deliver tangible results for some time past. Incidentally, climate change is one of the few areas where a continental position has been agreed. Mobilizing African solidarity and unity on any issues has never been easy given the continent's diverse national interests and agendas.

Conclusion and Policy Recommendations

Climate change continues to threaten both the present and future generations. Its growing worsening impacts in recent years have drawn remarkable global attention. One of such was the 2019 UN Climate Action Summit that drew participant from across the world and ended with a call for more global climate action, particularly on the part of state-stakeholders. Though a yearly event but in recognizing the increasing dangers of climate change and the need for accelerated global intervention, the COP 25 held in December same year (2019) concluded with a resolution calling for more climate ambition and action in line with the mitigation goal of the Paris Agreement and the UN 2030 Agenda for sustainable development.

On its part, the UN climate action summit succeeded in introducing a set of transformative initiatives for which state and nonstate stakeholders are responsible. But, it failed to secure concrete and immediate mitigation pledges from the world's top greenhouse gas (GHG) emitters. In the case of COP25, not much progress was achieved beyond the presentation of the Climate Ambition Alliance (CAA). The CAA in its presentation succeeded in merely reemphasizing the multistakeholder pledges made at the UN climate action summit. This chapter argues that the lack of complementarity between the identified two summits is not only worrisome, but also exacerbates the concern raised in the emissions gap report that the existing mitigation pledges otherwise known as NDCs, even if met, would not be enough to deliver the Paris goal. The urge for increased ambition pledges and climate action is therefore critical to closing the gap between the current assemblage of NDCs and the mitigation goal set by the Paris Agreement.

Furthermore, it is argued that though the separate but joint call for more climate ambition and action is a global question, it provides Africa and the EU in particular a fresh opportunity to deepen their existing climate change partnership. This chapter not only underscores the strategic positions and relevance of these two longstanding partners to the global fight against climate change, but it also highlights that the EU through the unveiling of the European Green Deal already set the pace for increased mitigation pledges involving world's leading emitters. It is yet to be seen though how much of the proposed mitigation objectives would be realized ultimately.

Going forward, this chapter recommends that Africa–EU climate change policies be aligned with the UN transformative initiatives on the global climate action, particularly the African components. Besides, there is the need for a more assertive and balanced Africa–EU relation, particularly in the context of climate change. Such relation should be based on common climate agendas, with clearly defined priorities and action plans including a joint implementation framework that not only support the transformative initiatives but also consistent with the mitigation goal of the Paris Agreement. Africa–EU climate change partnership should be adjusted to complement the efforts of the UN steering committee on global climate action, especially in terms of providing strategic guidance and monitoring of the implementation of the global transformative initiatives in Africa. While the EU is encouraged to continue to provide both financial and technical supports to African countries to enable them fulfill their pledges relative to the global climate action, there is need for Africa to also look inward for solutions. More focus should be directed at unveiling regional-based solutions to the climate change challenges facing the African continent beyond the UNFCCC framework.

Lastly, as the mitigation ambition proposed in the framework of the Paris Agreement is far from being achieved and that the full implementation of the Paris Agreement has yet to be concretized, because efforts are always underway at the annual COP to finalize its operational guidelines; more plausible concepts such as a truly multilateral partnership which involve collective actions and shared responsibilities in all aspects should be considered to have a good Africa–EU partnership on climate change in light of the call for more climate action.

References

Amusan L, Olutola O (2016) Addressing Climate Change in Southern Africa: Any Role for South Africa in the Post-Paris Agreement? India Quarterly 72(4):395–409. https://doi.org/10.1177/0974928416671592

AUC (African Union Commission) (2015) Agenda 2063: the Africa we want. Popular version. AUC, Addis Ababa

Bach D (2010) The EU's 'strategic partnership' with Africa: Model or Placebo? GARNET Working Chapter No. 80/10, September

Bilski A (2018) Africa's great green wall: a work in progress. Landscape News, August

Boyd R, Turner JC, Ward B (2015) Intended nationally determined contributions: what are the implications for greenhouse gas emissions in 2030? Policy Chapter, October. ESRC Centre for Climate Change Economics and Policy and Grantham Research Institute on Climate Change and the Environment

Burchill S (2005) Liberalism. In: Burchill S, Linklater A, Devetak R, Donnelly J, Paterson M, Reus-Smit C, True J (eds) Theories of international relations, 3rd edn. Palgrave Macmillan, Hampshire and New York, pp 55–83

CarbonBrief (2019) COP 25: key outcomes agreed at the UN climate talks in Madrid. 15 December. Retrieved from https://www.carbonbrief.org/cop25-key-outcomes-agreed-at-the-un-climate-talks-in-madrid

Claeys G, Tagliapietra S, Zachmann G (2019) How to make the European green Deal work. Policy contribution issue no. 13, November. Bruegel, Brussels

EC (European Commission) (2015) African Union Commission and European Commission meet to bring new impetus to the EU-Africa partnership. Press release, 21 April. EC, Brussels

EC (European Commission) (2019) The European Green Deal. Communication from the Commission to The European Parliament, The European Council, The Council, The European Economic and Social Committee and The Committee of the Regions. COM (2019) 640 final. EC, Brussels

EU (European Union) (2007) The Africa-EU strategic partnership: a joint Africa-EU strategy. 16344/07 (Presse 291), 9 December. Council of the European Union, Brussels

EU (European Union) (2013–2019) The partnership and joint Africa-EU strategy. Retrieved fromhttps://www.africa-eu-partnership.org/en/partnership-and-joint-africa-eu-strategy

EU (European Union) (2014) The Africa-EU partnership: 2 unions, 1 vision. Summit edition. Publications Office of the European Union, Luxemburg

GCCA (Global Climate Change Alliance) (2018) Our programs. Retrieved from http://www.gcca.eu/programmes

Globalist (2018) The scale of G20 Greenhouse Gas Emissions: can the G20 summit help advance international cooperation on the environment? 2 December. Retrieved from https://www.theglobalist.com/the-scale-of-g20-greenhouse-gas-emissions/

Grieco JM (1988) Anarchy and the limits of cooperation: a realist critique of the newest liberal institutionalism. International Organisation 42(3):485–507

IPCC (Intergovernmental Panel on Climate Change) (2018) Global warming of 1.5°C: an IPCC special report on the impacts of global warming of 1.5°C above pre-industrial levels and related global greenhouse gas emission pathways, in the context of strengthening the global response to the threat of climate change, sustainable development, and efforts to eradicate poverty. IPCC, Switzerland

Keohane RO (1984) After hegemony: cooperation and discord in the world political economy. Princeton University Press, Princeton

Keohane RO, Martin LL (1995) The promise of institutionalist theory. Int Secur 20(1):39–51. https://doi.org/10.2307/2539214

Keohane RO, Nye JS Jr (1987) Power and interdependence revisited. Int Organ 41(4):725–753

Khadiagala GM (2018) Europe-African relations in the era of uncertainty. In: Nagar D, Mutasa C (eds) Africa and the world: bilateral and multilateral international diplomacy. Palgrave Macmillan, Chaim, pp 433–453

Krasner SD (1983) Structural causes and regime consequences: regimes as intervening variables. In: Krasner SD (ed) International regimes. Cornell University Press, Ithaca and London, pp 1–21

Lawrence P, Wong D (2017) Soft law in the Paris climate agreement: strength or weakness? Rev Eur Comparat Int Environ Law 26(3):276–286. https://doi.org/10.1111/reel.12210

Martin L (2007) Neoliberalism. In: Dunne T, Kurki M, Smith S (eds) International relations theories: discipline and diversity. Oxford University Press, Oxford, pp 109–126

Miola A, Papadimitriou E, Mandrici A, McCormick N, Gobron N (2015) Index for the EU global climate change alliance plus flagship initiative. Publications Office of the European Union, Luxembourg

Moncel R, van Asselt H (2012) All hands on deck! Mobilising climate change action beyond the UNFCCC. Rev Eur Comparat Int Environ Law 21(3):163–176. https://doi.org/10.1111/reel.12011

NEPAD (New Partnership for Africa's Development) (2019) TerrAfrica. African Union Development Agency/NEPAD, Midrand. Retrieved from https://www.nepad.org/programme/terrafrica

Olutola O (2020) U.S. withdrawal from the Paris agreement: implications for climate finance in Africa. Africa Review 12(1):18–36. https://doi.org/10.1080/09744053.2019.1685334

Schleussner C, Rogelj J, Schaeffer M et al (2016) Science and policy characteristics of the Paris agreement temperature goal. Nat Clim Chang 6:827–835. https://doi.org/10.1038/nclimate3096

Tondel F, Knaepen H, van Wyk L (2015) Africa and Europe combating climate change: towards a common agenda in 2015. Discussion chapter no. 177, May. European Centre for Development Policy Management, Maastricht

UN (United Nations) (2019) Report of the Secretary-General on the 2019 Climate Action Summit and the way forward in 2020. Retrieved from https://www.un.org/en/climatechange/assets/pdf/cas_report_11_dec.pdf

UNCC (United Nations Climate Change) (2019a) Climate action and support trends: based on national reports submitted to the UNFCCC secretariat under the current reporting framework. UNCC Secretariat, Bonn

UNCC (United Nations Climate Change) (2019b) Climate ambition alliance: nations renew their push to upscale action by 2020 and achieve net zero CO2 emissions by 2050. External Press Release, 11 December. Retrieved from https://unfccc.int/news/climate-ambition-alliance-nations-renew-their-push-to-upscale-action-by-2020-and-achieve-net-zero

UNECA (United Nations Economic Commission for Africa) (2014) ClimDev-Africa annual report. 1 January – 31 December. UNECA, Addis Ababa

UNEP (United Nations Environment Programme) (2019) Emissions gap report 2019. UNEP, Nairobi. Retrieved from https://unepdtu.org/wp-content/uploads/2019/12/egr-2019.pdf

United Nations Framework Convention on Climate Change (UNFCCC) (2015) Conference of the parties: report of the conference of the parties on its twenty-first session. Held in Paris from 30 November to 11 December. FCCC/CP/2015/L.9/Rev.1. Retrieved from http://unfccc.int/resource/docs/2010/cop16/eng/07a01.pdf

Urpelainen J, van de Graaf T (2018) United States non-cooperation and the Paris agreement. Clim Pol 18(7):839–851. https://doi.org/10.1080/14693062.2017.1406843

5

Digital Platforms in Climate Information Service Delivery for Farming in Ghana

Rebecca Sarku, Divine Odame Appiah, Prosper Adiku, Rahinatu Sidiki Alare and Senyo Dotsey

Contents

R. Sarku (✉)
University for Development Studies, Tamale, Ghana

D. O. Appiah
Environmental Management Practice Research Unit, Department of Geography and Rural Development, Faculty of Social Sciences, Kwame Nkrumah University of Science and Technology, Kumasi, Ghana
e-mail: doappiah.cass@knust.edu.gh

P. Adiku
Institute for Environment and Sanitation Studies, College of Basic and Applied Sciences, University of Ghana, Accra, Ghana

R. S. Alare
Faculty of Earth and Environmental Sciences, Department of Environmental Sciences, C.K. Tedam University of Technology and Applied Sciences, Navrongo, Ghana

S. Dotsey
Urban Studies and Regional Science, Gran Sasso Science Institute, L'Aquila, Italy

Abstract

Phone-based applications, Internet connectivity, and big data are enabling climate change adaptations. From ICT for development and agriculture perspectives, great interest exists in how digital platforms support climate information provision for smallholder farmers in Africa. The vast majority of these platforms both private and public are for delivering climate information services and for data collection. The sheer number of digital platforms in the climate information sector has created a complex information landscape for potential information users, with platforms differing in information type, technology, geographic coverage, and financing structures and infrastructure. This chapter mapped the existing climate information services and examined their impact on policy and practices in smallholder farming development in Africa, with a focus on Ghana. Specifically, the chapter provides highlights of digital platforms available to smallholder farmers and agricultural extension agents, analyzes the public and/or private governance arrangements that underpin the implementation of digital climate information delivery, and assesses the potential of these platforms in scaling up the use of climate information. The chapter contributes to understanding the dynamics of climate information delivery with digital tools in Africa, and suggests a future research agenda.

Keywords

Climate information services · Digital platforms · Smallholder farming · Ghana · Africa

Introduction

Climate change and variability are complex problems affecting different geographical regions, people, and socio-ecological systems. Africa is likely to be the most adversely affected due to its overdependence on climate-sensitive sectors for development. For instance, studies have shown that Ghana is highly prone to drought; temperatures are projected to increase at the rates of 3.8% (1.02 °C), 5.6% (1.5 °C), and 6.9% (1.8 °C) for the near future (2040), mid-future (2060), and far future (2080), respectively, and rainfall is expected to become more erratic (Dumenu and Obeng 2016). The changing climate conditions and the projections of increased temperature and precipitations have implications for smallholder farming and household food security.

The Intergovernmental Panel on Climate Change (IPCC), in recognition of the heightened climate impacts, vulnerabilities, and need for adaptation, encourages the adoption of a cross-sectoral and integrated approach to the development and adoption of long-term and sustainable adaptation and mitigation strategies (IPCC 2014). Adaptation, along with mitigation efforts, is a complementary approach for reducing and managing the risks associated with climate change and sustainable development particularly in developing countries (Federspiel 2013; IPCC 2014). Countries are therefore expected to develop mechanisms (plans, programs, and policies) that enhance resilience and provide support for local practices for climate change adaptation (Dougill et al. 2017).

To support farmers to better adapt to climate change and variability, context-relevant, accurate, and timely climate information services (CIS) have become a relevant strategy (World Meteorological Organisation 2015). CIS range from climate data delivery, transformation of climate-related data together with context information into customized products such as projections, forecasts, trends, economic analyses, analysis on best practices, development and evaluation of solutions, user interactions, and capacity development (Vaughan and Dessai 2014; Vogel et al. 2019). Through the provision of CIS, farmers can make crucial decisions regarding when to plough, sow seed, apply fertilizer, and harvest, thus, reducing exposure to risks (Tall et al. 2014). It also provides actors such as agricultural extension agents, researchers, and policy-makers with the requisite knowledge to support farmers to adapt to climate change. Despite the relevance of CIS, smallholder farmers are unable to access it due to high illiteracy rates, technical information, financial constraints, sociocultural barriers, and infrastructural challenges (Singh et al. 2016, 2018). In this regard, the Global Framework for Climate Services (GFCS) has become a prominent mechanism to address these challenges (World Meteorological Organisation 2015). The framework recognizes the role of digital tools for the provision of tailored CIS to meet the needs of users including smallholder farmers (Singh et al. 2016). Digital tools include new ICTs such as smartphones and old ICTs such as radio and television (Asenso and Mekonnen 2012). The growing innovations associated with ICTs create new opportunities for information-sharing, alternative forms of connectivity, financing models, and governance arrangements (Bennett and Segerberg 2012; Karpouzoglou et al. 2016; Soma et al. 2016).

Despite the opportunities associated with the application of ICTs for the delivery of CIS in Africa, opinions abound regarding the extent to which smallholder farmers use ICT-delivered CIS.

Some of these opinions include:

- Old technologies such as radio, television, and farmers' networks remain the most relevant and cost-efficient ICT platforms for the delivery of CIS.
- ICTs remain underused despite the hype due to the lack of relevant content information, lack of or poor infrastructure, low affordability, low literacy, lack of access to ICTs, and absence of conducive social norms, such as trust.
- ICTs are social transformational tools which are enabling farmers to become co-producers of CIS rather than consumers.
- The delivery of ICT-based CIS results in changes in institutional logics in farming communities in the form of new interactions and information exchange and other forms of innovation intermediation.

The differences in the opinions on the application of ICTs for the delivery of CIS for smallholder farmers may be attributed to the relatively underdeveloped CIS sector in Africa (Graham et al. 2015; Hansen et al. 2019). This can also be attributed to the fact that studies on CIS use among smallholder farmers in Africa are often framed under themes, such as gender, accessibility, local content development, and local knowledge, while ICTs are mostly treated as sub-themes. Based on this gap, this chapter examines the application of digital tools for the delivery of CIS for smallholder farmers with a focus on Ghana. This chapter is guided by the following questions:

- Who are the actors of ICT-based CIS delivery?
- What digital tools are used for the delivery of CIS for smallholder farming in Ghana?
- Which category of farmers applies CIS?
- What financing modes and governance arrangements are used for the delivery of CIS?

The chapter contributes to the literature on climate change adaptation in two ways: First, it provides an overview of the landscape of digital tools that are being used to provide CIS to smallholder farmers in Ghana. Second, it provides insights and directions for further research on the delivery of CIS with digital tools in Ghana in support of adaptation. The chapter consists of an "Introduction", "Methodology", "Results", and "Conclusion" sections.

Methodology

To generate data for the chapter, a systematic literature review was carried out with three databases: Scopus, Google Scholar, and CAB Abstracts. Keyword searches of terms "Climate services" AND "Ghana," "Climate information services" AND "Ghana," "Weather information" AND "Ghana," and "Weather information services" AND "Ghana" were entered as queries. "Weather and climate information services" AND "Ghana," "Climate and weather enterprise" AND "Ghana," and "Agrometeorological services AND Ghana" were conducted. Other queries include "Hydroclimatic information," "Weather information services," "Agroclimatic information," "Services," and "Information." Articles that had the abovementioned key terms in their title or abstract were selected through a process of "abstract sifting" to help align with the language, concepts, and subject matter. Since CIS is still in its infancy and now gaining attention in the literature in Africa, limited literature was generated for the period 2000 to 2020. Hence, the snowball strategy was used to expand the scope of the literature by identifying additional literature from the selected peer-reviewed journals. Other literature which had combined themes on CIS and agricultural extension or had a regional scope on Africa with case studies on Ghana were also selected.

Additionally, desktop search was conducted by reviewing the websites of public, nongovernmental organizations (NGOs) and business organization that provide CIS in Ghana including ESOKO; MFarm; Farmerline; e-agricultural platform; Climate Change, Agriculture and Food Security (CCAFS); African Cashew Initiative, Ignitia, Technical Centre for Agricultural and Rural Cooperation ACP-EU (CTA); US Agency for International Development (USAID); and Agricultural Cooperative Development International and Volunteers in Overseas Cooperative Assistance (ACDI/VOCA). Relevant policy documents, thesis, and gray literature were also reviewed to provide information for the analysis. The limitation of the method used for the selection of literature for the chapter is the application of only three databases. Also, new publications could have been made after the literature search was conducted. The intertwining nature of CIS, ICTs, and other broad themes could have resulted in the omission of certain literature that might have relevant information on the application of ICTs for smallholder farming in Ghana.

The content analysis of the selected literature was guided by the research question and core themes: ICTs, digital tools, CIS, smallholder farming, and Ghana. Additionally, the content of the literature was analyzed based on the geographical settings, governance arrangements, geographic scope, type of digital tools or products, content of the digital messages, stages in their development, targets in the farming sector, and organizations providing CIS with ICTs. Here, the content analysis helped categorize and structure the sub-themes in order to select new themes emerging from the literature.

Results

Actors Involved in the Provision of CIS with ICTs

The actors providing CIS with ICTs to smallholder farmers were grouped broadly since a more comprehensive analysis indicating all actors is beyond the scope and objective of this chapter.

Actors who collected weather data with ICTs include the Ghana Meteorological Agency (GMet). At the international level, international weather organizations (National Oceanic and Atmospheric Administration (NOAA) and European Organisation for the Exploitation of Meteorological Satellites (EUMETSAT)), the European Centre for Medium-Range Weather Forecasts (ECMWF), the National Aeronautics and Space Administration (NASA), UKMet, and the World Meteorological Organization also provide weather data (Mills et al. 2016; Usher et al. 2018). Farmers also constitute CIS data providers as they also generate information on local weather indicators in their communities for co-creating weather information with formal scientific models (see Kadi et al. 2011; Nyadzi 2020).

Intermediary CIS providers add value to the weather data. They include business weather data providers, like aWhere, Satelligence, and Weather Impact, mobile telecommunication network providers, and ICT platform providers (e.g., Prepeez) (Singh et al. 2018), while some knowledge institutions also play this role. Some actors double as intermediary CIS providers and also deliver the information directly to farmers with their ICT platforms, e.g., Farm Radio International, MFarm, Esoko, Farmerline, and VOTO Mobile among others.

The delivery of CIS information for farmers is carried out by a variety of actors from the government agencies, e.g., GMet, GTV, National Disaster Management Organization, and regional public radio channels (Adjin-Tettey 2013). International development partners in collaboration with ICT companies also provide CIS to farmers (Mohammed 2018). Businesses, commercial radio channels, and agri-input companies are also involved in the delivery of CIS to farmers. Other actors include processors, input suppliers, aggregators, exporters, agricultural service providers, and NGOs. They focus on providing complex interventions (e.g., demonstrations, visual aids, agronomic inputs, logistics, and resources) to ensure farmers' access to information (Slavova and Karanasios 2018).

Digital Platforms for the Delivery of CIS to Smallholder Farmers

This section is divided into three subsections.

Data Collection and Production of CIS

Few studies provide information on the application of ICTs for collecting data and producing CIS for the smallholder farming sector. Even so, some studies which captured data collection at the farming community level were identified. These include ground-level measurements and nowcasts that are derived from ICT-enabled platforms such as algorithms, automatic weather stations, weather models and historical observational weather data, automatic rain gauge, sensors, data logger, and ground-level radio with the aid of mobile telecommunication network radar (Gotamey et al. 2018; Usher et al. 2018; Nyadzi 2020). Global Positioning Systems (GPS) are also used to pick the location of communities, and some of the abovementioned technologies are triangulated to provide location-specific data. The mobile phone or smart technologies are also used to transmit data from weather stations in different locations to weather information producers, although carried out on a trial basis (Akudbillah 2017; Usher et al. 2018). In the case of the actual provision of weather data, the technologies are mostly based outside Ghana, and so the role of the Internet becomes relevant for data transmission (Chudaska 2018). For example, supercomputers and cell tower-mounted or co-located automatic weather stations capitalize on the increasing availability of secure, powered, connected telecommunication infrastructure in Ghana. Smartphones are used to collect personal data, type of crops cultivated, and location with Geographic Information Systems (GISs), and the data is stored to a server. Also, some organizations have researched the production of CIS with ICTs, such as chip sensors, in the northern part of the country for monitoring temperature and other weather indicators in local communities (Mills et al. 2016; Usher et al. 2018). The collection of local weather indicators with a smartphone for co-producing location-specific CIS has also been carried out on a trial basis (Nyadzi 2020; Sarku et al. forthcoming).

Data Management and Analysis

CIS information providers usually have supercomputers, data processing, analytical models, sophisticated algorithm proprietary data sets, or multilayer ICT architecture with a system of modular components (functionalities and interfaces) that communicates with a central cloud application, which includes a central database (see Fig. 1) to interpret and convert raw data into accurate forecasts. For example, Ignitia uses supercomputers to provide real-time weather forecasts to farmers. The geospatial data collected from many locations over time are collated in a central database and then interpreted by experts (Chudaska 2018). Application Programming Interfaces (APIs) are used to deliver real-time CIS data to systems and businesses (e.g., aWhere and Farmerline). Cloud computing is used to limit the impact of unstable power and air conditioning systems in computer rooms (e.g., aWhere and Ignitia). CIS bundling is one of the key marketing strategies of market-led ICT platform information providers. It involves the collection of GPS coordinates and profiles on farmers' socioeconomic characteristics to provide value-added services such as agronomic advisory (Etwire et al. 2017; Partey et al. 2020). Some pilot projects indicate the use of different ICTs to integrate the weather data with soil and crop advice and make suggestions for minimizing losses and optimizing inputs (Eitzingera et al. 2019). Decision-support systems such as RainCast application have data source that uses OpenWeatherMap with the aid of an API to extract weather data (Omoine et al. 2013). From the API, the app retrieves temperature or weather conditions for the day and the time the condition is expected to occur for as much as 5 days ahead (Dinku et al. 2018; Gotamey et al. 2018). Mobile telecommunication networks also help to locate users in a variety of ways through network-based call detail records (CDRs), triangulation via LBS, and user registration, which enable the creation and delivery of highly localized, farm-level forecasts based on user location.

Information Delivery

To enable the delivery of CIS to farmers, the focus is mostly on the medium of information delivery due to its influence on the usability of the information for decision-making in farming (Dilling and Lemos 2011; Lemos et al. 2012). Often a distinction is made in the literature between existing and emerging technologies as "old" and "new" ICTs, respectively, based on different factors. From the social perspective, ICTs are classified as new, due to the ability to mostly complement or spur real-life interactions, and two-way communication (Materia et al. 2015; Sulaiman et al. 2012; Slavova and Karanasios 2018). They also enable alternative forms of connectivity and virtual communities for different actors (Bennett and Segerberg 2012; Cieslik et al. 2018). Another factor that is used to distinguish ICTs as either old or new is based on the capacity of the technology to collect data, process, and deliver information within a short period across space and time (Asenso-Okyere and Mekonnen 2012; Munthali et al. 2018). Also, ICTs are classified as new based on the innovations attached to the technologies such as the ability to provide mobile push-and-pull services (e.g., social media) (Bell 2015; Barber et

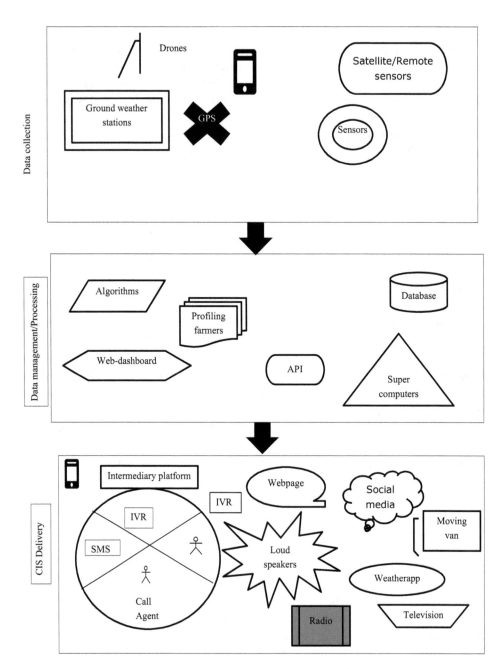

Fig. 1 Flow chart indicating how digital platforms are used for the delivery of climate information services

al. 2016; Suchiradipta and Saravanan 2016). In some instances, new technologies are identified to support environmental monitoring through the involvement of citizens in data collection (Buytaert et al. 2014; Karpouzoglou et al. 2016; Nyadzi 2020). With regard to the governance or institutional perspective, old ICTs have remained large under the control of state organizations, while new ICTs seem uncontrollable and have transformative powers due to the numerous actors who use it to provide information across space and time (Mol 2008; Soma et al. 2016). In reference to these factors, a plethora of new ICTs is identified for the delivery of CIS for farming (Akudugu et al. 2012; Munthali et al. 2018).

New ICT

Mobile phones for CIS delivery are increasing in Ghana due to its different functionalities. As tools for oral interpersonal communication, mobile phones have a high level of ownership among farmers (Anuga et al. 2019). Farmers stressed the importance of mobile phones as tools for accessing formal and informal networks. For instance, they are used to contact agricultural extension agents (e.g., dealing with pests) or reach out to fellow farmers for assistance (Alhassan et al. 2013). The are several advantages associated with the use of a mobile phone for the delivery of CIS (see Table 1). The functions of mobile phones used by RainCast for the delivery of CIS include a selection of options for language, location, type of forecast required, and other information related to farming (Omoine et al. 2013).

Interactive voice response (IVR) is a voice-based channel of communication; others include call centers and voicemail (Zougmoré et al. 2018). The e-extension platform for the Ministry of Food and Agriculture (MoFA) has a mobile application called Farmer Direct which allows farmers to call in and receive pre-recorded information (Modernizing Extension and Advisory Services (MEAS) 2012; Munthali et al. 2018).

The voice/audio and video messages are in the form of pre-recorded information in the local languages to meet the needs of different categories of farmers. The voice forum is a feature that allows farmers to ask questions by calling a toll-free helpline with a short code. Agricultural extension agents can answer questions via a web interface, and answers are sent to farmers as voice SMS (Slavova and Karanasios 2018).

Short message services (SMS) are short written text with description on weather conditions and other agricultural information on how to tackle pests or diseases, agricultural techniques, optimum planting times, available subsidies, and weather forecasts, local fairs, and crop prices (Aker 2011). For example, Ignitia Ghana partners with ACDI/VOCA to deliver weather forecast through SMS to farmers who are part of the Agricultural Development and Value Chain Enhancement (ADVANCE) II project. Automated SMS alerts are also used for the delivery of CIS (Mohammed 2018).

Weather apps such as AccuWeather, Rainsat, and others are installed on smartphones to provide a daily forecast. Some of these apps provide a forecast for 48 h or the entire week. Some weather apps are available to specific network subscribers (Caine et al. 2015). The smartphone has also enabled user-specific weather apps with an interface designed for several functionalities and applications for farmers (Gotamey et al. 2018).

Mobile-based-only platforms include MTN and Vodafone telecommunication network scribe to access CIS with SMS and IVR. Vodafone farmer clubs have

Table 1 Overview of ICT platforms for information delivery to farmers

Categories	Type of ICT platform	Application during the farming seasons	Advantages	Challenges
Voice-based	Mobile phone	All-year usage	• Social identities through personalized ringtones • More durable and portable • Can be used in numerous ways due to the multiplicity of functions for illiterate farmers • Missed calls can be traced • Cheap technology • Allows for two-way interaction with other farmers or agricultural extension agents • Two-way flow of information provides opportunities for information providers to generate feedback	• Access to messages not free • Illiterate farmers cannot access messages delivered in the English language • Novice users cannot trace or navigate through the phone to trace missed calls • Most existing phone-based services are not free, and this raises the issue of information asymmetry where farmers who can pay access to information • Loss of phone results in loss of messages and cost implications
	IVR (interactive voice response)	Used only during the farming season	• Available on feature phones • Interaction with information provider is possible • Accessible for illiterate users • Has the potential to reach more people • Richer content • Cost-effective alternative to access information	• More expensive than a text message • Require training to use
	Television	All-year usage	• Free access • Uses images and symbols • Oral presentation	• No interaction with viewers • Access affected by the unavailability of electricity • Weak network signals and poor reception • Communication of CIS is in English • TV ownership not common among farming households • Unfamiliarity with symbols depicting weather conditions such as rainfall or sunshine

| Interactive community and commercial radio | Mainly at the onset of the farming season | • Can be powered by batteries or solar panels
• Free access to information
• Exist in portable forms and movable
• Multiplicity of radio channels in a specific area with wide frequency coverage
• Presentation of information in (multiple) local languages
• Timely broadcast of CIS programs
• Satisfies the needs of all listeners including illiterates
• Engages different panelists with different knowledge levels
• Formation of listenership groups in communities
• Integrated with phone-in activity to enable listeners share experience, challenges, and questions
• Discussions bridge gap in communication of uncertainties in the forecast
• Uses voice prompts, vox pops, beep calls, Beep4Weather system, and pre-recorded information
• Values farmers' local knowledge, cultural beliefs, and attitudes
• Local contents embedded in discussions build trust in CIS
• Level of dialogue is synonymous to face-to-face methods
• Integrates a variety of ICTs into the programs, e.g., phone-in | • Provides less opportunity for in-depth discussion and interaction with panelists
• Phone-in sessions during the in-studio discussion are limited
• Schedule programs and limited time allotted for CIS programs throughout the week, e.g., 1 h per week
• Program formulation can be top-down
• CIS programs run mostly at the onset of the season
• Possible altering of information in the process of translating from English to the local languages
• Broadcasts are fleeting; one either hears them when they are broadcast, or they are missed
• Limited ways of accessing missed programs unless they are recorded and played back
• Affordability of batteries or solar-powered radio sets
• Limited in facilitating learning-by-seeing or visual presentations |

(continued)

Table 1 (continued)

Categories	Type of ICT platform	Application during the farming seasons	Advantages	Challenges
	Video (YouTube)	All-year usage	• Facilitate better comprehension of complex ideas, specifically for illiterate farmers • Access to mobile internet opens significant new possibilities for knowledge-sharing and exchange with social networks • Allows for visual display of data and discussions	• Not easily applicable in all devices • Poor Internet connectivity affects its use
Text-based messaging	SMS (short message service)	All-year usage	• Available on all mobile phones • Low-cost message • Low technical setup • Can be stored for later viewing or forwarded • Information is straightforward • Easily accessible on any mobile phone • It is readily available	• Deductions debited from call credit • SMS nuisance due to numerous messages from telecommunication networks • Requires two-sim phones to subscribe if farmers are subscribers of other networks • Information delivered in English • Inability to read information deters some farmers from accessing information • No graphic communication • Messages are sometimes too short to allow adequate comprehension of information • Limited communication of uncertainty • Scheduled delivery of SMS at fixed times • No provision of outlook forecast for the day • Lack of interactions
	Mobile network based on platform	Used only during the farming season	• Service providers enable free calls between subscribers • Access to an expert helpline (Vodafone 2015) • Simple and comprehensive messages • Voice SMS in 13 local languages • Provision of a variety of temporal CIS	• Disadvantages same as SMS

Platform	Usage	Features	Challenges
Mobile weather apps for smartphones	Irregular use during the farming season	• Almost unlimited • Informative and intuitive • Visualizations are possible • Symbols exist to illustrate CIS • Visualization capabilities in mobile apps help to overcome the challenge of limited literacy rate in rural communities	• Low level of smartphone ownership • Mobile data expensive • Unavailability of the Internet • Inability to use a smartphone • Complexity attached to navigating apps • Inaccessibility to the Internet to enable the use of online mobile apps • Many farmers are smartphone phobia due to language and literacy barrier
Smartphones	Irregular use during the farming season	• Smartphone application is simple and optimized for fieldwork usage • Variety of functions, e.g., social media, registration of farmers, pick GPS points, apps • Can be used for offline data collection, and data can be uploaded upon the availability of the Internet	• Difficulty in swiping sensitive screen, switching data, and renewal of Internet data and other more sophisticated functions • High cost of Internet data • Inaccessible Internet in most locations • Technology not coded in local language • Keyboard does not have alphabets for local languages
Website	Irregular use during the farming season	• Website that provides weather fore for 24 h • Focused on all regions in Ghana	• Offer limited opportunity for interactions • Limited Internet access affects its use in many rural communities • Farmers will find it difficult to read and interpret information by themselves because of limited literacy
Newspaper and bulletins	Mostly used at the onset of the farming season or for outlook forecast	• Written text suitable for literate farmers	• Constrained by low literacy rates • Scheduled publication deliveries • Available at certain periods of the year • Accessibility to a few farmers and in urban areas

(continued)

Table 1 (continued)

Categories	Type of ICT platform	Application during the farming seasons	Advantages	Challenges
Intermediary platforms	Platform intermediaries	Used only during the farming season	• Play intermediation roles • Engages farmers to identify their tacit knowledge and integration with scientific CIS • Foster continuous information flows and trust building • Collection of databases for information on farmers • Aids in scaling up access to CIS through the combination of SMS, call-in services, rural radio, agricultural extension or field agents, farmer organizations, and social networks • Offers learning opportunities in broader networks	• Inadequate experts to play the intermediation roles • Logistical constraints in staff deployment
	Call center	Used only during the farming season	• Provision of expert advice • Voice-based calls • Access the helpline by dialing a short code at a regular call rate • Provides tailored and interactive solutions • Offers timely responses to all category of farmers • Provides a mechanism to receive feedback on the service • Offers services in a variety of languages • Builds trusts in CIS • Helps in contextualizing the information and discuss risks	• Unavailable 24/7 • Cost of call credit for farmers

Informal: van, loudspeakers, community information center	Mostly used at the onset of the farming season	• Facilitates intensive face-to-face interaction • Can target a wide audience • Information presented in the local language • Relatively low cost	• Information not frequent • CIS available at a specific period of the season • Other social issues prioritized above CIS
Social media/ weather app	Irregular use during the farming season	• Combines text, images, animations, symbols, and videos • Provides timely delivery of climate information • Low-cost information delivery • Enables information to travel easily • Attractive forms of communication • Can provide outlook forecast during the day	• Requires advance ICTs such as apps and smartphones • Not accessible to all farmers • Unavailability of the Internet and other infrastructures can be challenging • Inability or difficulty in using ICTs, e.g., mobile phone, smartphone • Farmers' inability to navigate several ICTs and read messages • Schedule delivery • Little local content and guide on the application of information

Source: Authors' construct based on literature review

been created as part of the setup enabling the provision of information on nutrition, weather information, and market prices for eight regions in Ghana: Eastern, Western, Ashanti, Central, Northern, Volta, Brong-Ahafo, and Greater Accra (Aker et al. 2018). Vodafone has partnered with ESOKO whereby subscribers to the Vodafone network are connected to input suppliers, aggregators, extension services, and other value chain actors (Nyarko et al. 2013; Duncombe 2016).

Social media platforms such as YouTube, WhatsApp, and Facebook play a role in the delivery of CIS by GMet. YouTube video showcases a 2-minute video on nationwide weather forecasts. On GMet's Facebook page, there are flyers of impending weather conditions for the day or the bulletins to provide an outlook for changes in weather conditions on specific days. The role of WhatsApp is to provide a real-time forecast on an impending storm spotted on satellite images. At the onset of the rainy season, GMet also disseminates its seasonal forecast through YouTube or Twitter.

The website is another channel for the delivery of CIS but mostly it is not tailored for farming. Websites provide general CIS to the public on how to prepare for the dry season or impending floods (Anuga et al. 2019). For instance, during the minor season in 2019, GMet used its website to provide an overview of the weather conditions for the seasons.

Concerning print media such as the newspaper and seasonal forecast bulletins, information on the weather forecasts is communicated via state-owned electronic media and published bulletins from Ghana Media Association (GMA), Ghana News Agency, and the *Daily Graphic*. Some of the outlooks for the rainy season are also published in newspapers (e.g., daily graphic), and it is usually at the onset of the main season.

Intermediary platform is also used for the provision of CIS for farming. Intermediary platform consist of farmer audio library, IVR where a farmer can call a specific toll-free line and is taken through the procedures for the desired CIS in their local languages, and agricultural extension agents or field agents who are also equipped with smartphones or tablets to provide a two-way information. It also consists of a call center equipped with agents and a special telephone number (MEAS 2012, Agyekumhene et al. 2018; Munthali et al. 2018; Slavova and Karanasios 2018). With intermediary platforms, field agents or agricultural extension agents serve as mediators between the ICT platforms and farmers. These intermediaries ("infomediaries") may reside within the farming communities, access information on behalf of farmers, and discuss it with them. In some of the case studies, intermediaries are lead farmers and community knowledge brokers. Examples of intermediary platforms include ESOKO, e-agricultural platform, Smartex, and GeoFarmer (Munthali et al. 2018; Quaye et al. 2017).

Old ICTs

Despite being conventional, old ICTs are also applied in innovative ways to suit the needs of rural farming communities, especially where the use of old ICTs dominates and is intertwined with the social networks and institutional logic of farmers. This contributes significantly to enhancing the spread of information in the farming communities (Duncombe 2016). The chapter captured the nuanced interplay of innovation with old ICTs.

Radio is one of the conventional digital means of communication that is still relevant in the delivery of CIS to farmers in Ghana (Sam and Dzandu 2016; Nyantakyi-Frimpong 2019b). The use of radio is popular in CIS delivery for several reasons (see Table 1). Community radios, such as Simli Radio at Dalung and Radio Ada at Big Ada, create interactive programs which feature community discussions, interviews, live panel discussions, and call-in shows in local languages. CIS is broadcasted on community radios at specific days in the week when farmers are available, and special signature tunes to draw the attention of listeners (Asenso-Okyere and Mekonnen 2012; Chapman et al. 2003; Nyamekye 2020). Programs are designed and periodically updated throughout the year in response to the types of CIS and agricultural information that are needed (Perkins et al. 2015). Sometimes, farmers are consulted directly during radio programs for relevant data on local indicators for the onset of the season, and comparisons are made with a scientific forecast in a panel discussion to further interpret information (Sam and Dzandu 2016). At a point in the programs, listeners are permitted to call in to interact with panelists, ask questions, and share problems and experiences. At times drama and pre-recorded programs are aired when a panel discussion is not carried out. Commercial radios also provide CIS (Chudaska 2018; Slavova and Karanasios 2018; Nyamekye 2020).

Government- and private-based television channels are other means through which farmers receive CIS, for example, GTV and TV3 (Adjin-Tettey 2013; Anuga et al. 2019). A study also indicated that some farmers also obtain weather forecasts from international television stations such as Aljazeera, CNN, and BBC (Chudaska 2018).

Informal CIS delivery channels have been shown to be preferred by farmers due to oral communication and personal conversation that are mostly practiced in farming communities (Drafor and Atta-Agyepong 2005; Drafor 2016). Farmers still prefer CIS to be delivered by agricultural extension agents, workshops, village knowledge centers, and mobile vans and pamphlets produced by NGOs (Nyantakyi-Frimpong 2019b). Traditional information mediums like drums, loudspeakers, public address systems, gong-gong beaters, town criers, churches, and the loudspeakers of mosques are still relevant (Padgham et al. 2013; Drafor 2016). Others include CIS delivery through Ghana Information Services Department's mobile vans (Anuga and Gordon 2016; Slavova and Karanasios 2018). The information received by farmers is then communicated to others in the farming community. In some communities, CIS is displayed on notice boards, while farmer-to-farmer or farmer cooperatives are also used to facilitate information delivery (Anuga et al. 2019).

Category of Farmers' Use of CIS Delivered Through ICTs

CIS is delivered to farmers with different socioeconomic backgrounds: gender, marital status, and level of education. Mostly, men are able to access CIS with a

variety of ICTs than women due to their ability to easily access mobile phones which
are the most common new ICTs in rural communities in Ghana (Partey et al. 2019).
Women's access to digital tools is limited to radio and mobile phones which are often
owned by their husbands. In some cases, women who had no formal education rely
on local knowledge (Chaudhury et al. 2012; Partey et al. 2019). Despite women's
use of the radio as the main ICT channel to receive CIS, they are unable to access
regular information due to their engagement in numerous household activities. Apart
from the ownership of the ICTs, men have a higher capacity to use and operate ICTs
especially the mobile phone than women without relying on others for assistance
(McOmber et al. 2013). This is attributed to the level of education between men and
women (Nyantakyi-Frimpong 2019b). The usage of mobile phones is limited to
simple handsets with the exception of a few educated and young farmers. Young
people (20–29 years) use new ICTs such as smartphones, apps, and social media.
Hence, they are able to access CIS with all forms of ICTs. Within the category of
men, young male farmers mostly received SMS and IVR (Partey et al. 2020).

Mostly, the beneficiaries of the CIS-funded projects select a household member or
recognized owner of household land to access the information with the assumption
that they would share the information with other members of their household
(Mohammed 2018). With the application of such criteria, women do not access the
information directly (Anaglo et al. 2014; Nyantakyi-Frimpong 2019a). Most women
are unable to access CIS with ICTs because of the domination of patriarchal norms
which promotes access and control of productive resources as well as decision-
making authority by men (Nyantakyi-Frimpong 2019a). Even if women are selected
to receive CIS with digital tools, the application of the criteria "one member per
household" results in the selection of an adult woman. This implies that within
polygamous households where seniority is a norm, it is the elderly women who gain
access to CIS.

The frequency of receipt of CIS delivered through ICTs range from daily, 48 h,
5 days, weekly, and monthly forecast. However, there are instances when lots of
pending or unopened messages are seen on farmers' phones, and they wait until their
literate relatives are around (Mohammed 2018). The reason for this is attributed to
low levels of literacy or inability to manipulate the phone (Alemna and Sam 2006).
The reliance on relatives who own ICTs to access CIS remains a challenge since their
absence could mean the inability to receive information.

Farmers have different types of information needs during each stage of the
farming season, and ICTs are used to meet these needs. Seasonal forecasts are CIS
provided to farmers at the onset of the farming season with interactive platforms,
such as radio and call center. They come with advice on how much rain to expect, the
length of the season, and the onset dates (Gbetibouo et al. 2017; Nyantakyi-
Frimpong 2019a). Additionally, SMS, IVR, bulletins, social media, websites, and
print media are also used to provide seasonal forecasts to support farmers' decision-
making on the selection of a variety of (specific) crops.

During the farming seasons, the provisions of local content- and location-specific
forecast with the IVR and SMS are mostly used. Also, the weather influences crop
pest and disease conditions during the farming seasons; therefore, intermediary

platforms such as call centers are used by farmers to generate information on the application of agrochemicals (Munthali et al. 2018). Some CIS have a short-term lifespan in terms of relevance to decision-making and therefore must be delivered within an appropriate timescale to farmers (daily up to 7 days) with different forms of ICTs (Zougmoré et al. 2018). In this regard, the intensity of the rains and the number of rainy days require the provision of daily forecasts with SMS, IVR, and radio to enable decision-making on when to weed and conduct other fieldwork (Partey et al. 2020). When floods or dry spell conditions are expected to occur, radio, SMS, social media, and intermediary platforms are used as the main channels to deliver outlook for the day or week (Stigter et al. 2013). Besides, ICTs help to contextualize CIS and integrate it into farmers' own schemes of experiences (Leeuwis 2004). Some of these CIS include information on the availability of transport and tractor services, credit options, input and market prices, and crop insurance during the farming season (Adiku et al. 2017; Anuga et al. 2019). For instance, the Vodaphone farmer club SMS provides a holistic CIS such as weather information, market prices, costs and availability of inputs (seed and fertilizer), and pest recommendations (Usher et al. 2018; ESOKO 2020). At the end of the farming season, SMS, IVR, and other informal channels are prominent means of delivering CIS such as a 10-day forecast for optimum harvesting period and processing of farm produce.

Type of Information Needed by Different Categories of Farmers

Farmers' information needs are associated with major decision-making for the farming season. The decision-making includes the type of crop and the variety of seed to select, land size and the number of farmland to cultivate, when to plough, when to sow seeds, when to transplant seedlings, when to do irrigation, when to apply fertilizer and(or) agrochemicals, and when to start harvesting, marketing, and processing (Sarku et al. 2020).

The scant literature on CIS indicates that gender plays a role in the information needs of different categories of farmers. Men have more CIS informational needs to mitigate climate risk compared to women. This is due to differences in preferences and vulnerabilities to climate-related risks, access to farm resources, land ownership, access to labor, access to information, and financial resources (McOmber et al. 2013). Jost et al. (2016) also indicated that limited access to finance, extension services, and farm resources affected women CIS needs. Furthermore, access to a mobile phone, access to irrigation, type of crop produce, and land ownership were influential in determining whether or not someone will use CIS (Partey et al. 2020). Table 2 shows an overview of the information needs for categories of farmers.

The first key decision for the farming season in relation to the application of CIS is the type of crop and the related variables that need to be cultivated. In most households, this decision is the responsibility of household heads who are mostly men. In some contexts, the household head and spouse (or spouses if polygamous) discuss privately and communicate with the rest of the family (Nyamekye 2020).

Table 2 Information needs of categories of farmers

Type of forecasts	Information needs	Categories of farmers			
		Young men	Young women	Elderly men	Elderly women
Seasonal forecast	Type of crop and the variety of seed to select			✓	✓
	Land size and number of farmland to cultivate			✓	✓
	When to plough	✓	✓		
Daily-weekly forecast	Sowing of seeds or transplanting of seedlings	✓		✓	
	When to apply fertilizer and other agrochemicals	✓		✓	
	When to weed	✓	✓	✓	✓
	When to start harvesting	✓	✓	✓	✓
	Marketing and processing				✓
Legend	✓	Applicable to specific category of farmers			

Source: Authors' construct based on literature review

This implies that young men's and women's information need concerning CIS is mostly dependent on the household heads (Nyantakyi-Frimpong 2019b).

Once farmers decide on the type and variety of crop(s) to cultivate, the next information needs is a forecast on the onset of the rainfall. This information is required to decide on when to prepare the land. Since land is usually prepared when soil moisture is assumed to be high, arrangement for land preparation with a tractor or manual labor becomes the responsibility of the male in the farming household (Chudaska 2018; Partey et al. 2020).

Sowing of seeds or transplanting of seedlings is carried by all categories of farmers after a rainfall. With this practice, CIS is needed by all categories of farmers. Additionally, men usually carry out the application of agrochemicals (e.g., application of fertilizer, weedicides, pesticides, and insecticides). Considering the cost of farm inputs and the availability of labor, CIS is needed by all categories of farmers to decide on when to apply agrochemicals (Sarku et al. 2020). The provision of a forecast on the amount or frequency of rainfall for the farming season is also required by all categories of farmers since it enables them to make arrangements for labor to augment household effort (Nyamekye 2020).

The study deduced that women's information needs to follow the same pattern as men yet it is tied to their relationship with men due to the unavailability of resources at their disposal. In instances where women are in charge of a household and need to take decisions concerning when to start farming, they often consult the men in the household or experienced male farmers in the communities (Nyantakyi-Frimpong 2019a, b). In some cases, women more often than not expect their husbands to provide them with information or determine the various aspect of the decisions for the farming season even when they operate different farms due to their husband's control over the household (Etwire et al. 2017; Partey et al. 2020). This finding adds evidence to support the growing call for the provision of CIS with ICTs which suit the needs of different categories of farmers.

Financing and Governance Arrangement for the Delivery of CIS with Digital Tools

CIS is a public good that can be accessible to all farmers to enable them to adapt to changing weather and climatic conditions. Yet the application of digital tools for the delivery of CIS has several costs associated with data collections, weather models, Internet data, and telecommunication networks. Therefore, CIS delivery in Ghana comes in the form of quasi-public good with some excludability as users pay for CIS to enable the recovery of cost (Naab et al. 2019). The cost of CIS delivery with ICTs brings to the fore discussions on farmers' willingness to pay for CIS delivered through ICTs (Quaye et al. 2017).

In this regard, two patterns can be identified: First, the majority of smallholder farmers recognize that the provision of CIS is a public good with the responsibility of the government to provide such services. Farmers who rely on free CIS delivered through the radio or television do not have access to tailored information, or they rely on their local knowledge in predicting the weather for farming (Nyantakyi-Frimpong 2013; Naab et al. 2019). Second, some farmers are willing to pay for CIS delivered through ICTs but within a minimum range (Acquah and Onumah 2011; Nantui et al. 2014). These categories of farmers have some socioeconomic characteristics such as age, sex, farm size, on-farm income, membership of a farmer-based organization, literacy level, productivity of crops in a particular season, and perception of climate change experience. Ownership and ability to use ICTs also influence farmers' decision to pay for seasonal CIS. Some studies have, however, indicated that male household heads are more willing to pay for CIS delivered through ICTs. Hence, farmers' willingness to pay for CIS delivered through ICTs is mixed and inconsistent (Adjabui 2018; Ibrahim et al. 2019). The financing mechanisms for the delivery of CIS by an organization involve several options as indicated in Table 3.

Linked to the financing model is the governance model which backs the provision of CIS with ICTs to farmers in Ghana. Although the governance model is not codified due to the absence of a CIS policy in Ghana (Naab et al. 2019), different governance modes are used for CIS delivery to farmers.

Donor or NGO Governance Approach

International donor agencies or organizations establish agricultural projects within specific agroecological zones for farmers. Some projects may be specific toward the provision of CIS, while others have different objectives (Asuru 2017). Numerous donor agencies or NGOs are operating with this governance model, and the provision of an exhaustive list is beyond the scope of this chapter. Few notable ones include ACDI/VOCA, TechnoServe, World Vision, FAO (Food and Agricultural Organization), USAID, Deutsche Gesellschaft für Internationale Zusammenarbeit (GIZ), International Fund for Agricultural Development (IFAD), CARE

Table 3 Financing of CIS with ICTs to farmers in Ghana

Category of CIS financiers	Funding mechanism
Government financing	Free provision by GMet and GTV through tax and sometimes product levies
Financing by agri-input companies, e.g., Wienco Ghana Limited	Provides embedded services by including the cost of CIS as part of the cost of the product
Commercial output buyers	Makes CIS freely available but incorporates the cost as part of the contract farming
Farmer-based associations, e.g., Cocoalink	Membership fees, donor subsidies, government subsidies, and contracts
Direct payment	Direct fees paid through farmers' call credit or airtime, e. g., Esoko
Funded projects	Donor contracts information providers to deliver CIS freely to target groups of farmers
Contract sale	Farmers' subscription is paid by insurance companies or agribusinesses
Business advertisement on CIS platforms	Advertisement of agricultural products linked to the provision of CIS
Minimum payment under public-private partnership	Subsidized payment is made by farmers under a project compared to those who are not part of any project

Source: Based on authors' literature review

International, OXFAM, and the Department for International Development (DfID) (McNamara et al. 2014). These NGOs partner mostly with market-led ICT platform information providers such as ESOKO, MFarm, Ignitia, Farmerline, and others to provide CIS to farmers. Examples include ADVANCE II, Smartex, African Cashew Initiative, and Farm Radio International's African Farm Radio Research Initiative. These projects mainly focused on specific areas such as the Guinea and Sudan savanna agroecological zone of Ghana due to the presence of donor activities in these areas (Naab et al. 2019).

Public-Private Partnership

The use of digital tools for the provision of CIS results in an interconnected network of activities between smallholder farmers and actors providing and mediating the process leading to the formation of public-private partnerships (PPP). The PPP governance mode of providing CIS for farmers involves NGOs, public sectors, mobile network telecommunication providers, market-led ICT platform information providers, and climate/weather model or satellite providers together with farming communities (Partey et al. 2019). With these arrangements, each actor provides certain resources for CIS delivery. Examples of PPP in the provision of CIS with ICTs include the CGIAR Research Programme on Climate Change, Agriculture and Food Security (CCAFS); businesses such as ESOKO, Vodafone, and Toto agric and aWhere; and public organizations that collaborate with GMet, the Council for Scientific and Industrial Research (CSIR), and MoFA to provide CIS to some farmers in the Upper West region under the Planting for Food and Jobs programs (Etwire et al. 2017; Partey et al. 2019, 2020).

Participatory or Co-production Approaches

ICTs enable cross production of knowledge from different sectors of society (Karpouzoglou et al. 2016; Vogel et al. 2019). In Ghana, ICTs are enabling this process through co-production of CIS with farmers' local knowledge and that of scientists. In some cases, farmers collect data on their local weather indicators, or they contribute their knowledge in several ways (Nyadzi 2020; Sarku et al., forthcoming). For example, Participatory Integrated Climate Services for Agriculture (PICSA) is an approach that has involved over 5000 farmers in Northern Ghana by using both historical climate information and forecasts to support farmers' decision-making with participatory decision-making tools that suit their contexts (Etwire et al. 2017; Clarksona et al. 2019).

Farmer Organization Provision

Another governance approach is the provision of CIS by producer cooperatives/associations. These organizations deliver CIS to farmers in the group. Payment for the CIS is drawn from the coffers of the association or through donor support. Examples of such governance models are used by Kuapa Kokoo cooperatives, the Africa Cashew Initiative, and Cocoalink (MEAS 2012).

Business Model

Business companies use a wide variety of business models to deliver CIS, and no single model is dominant. Six different business models were identified:

- Business-to-business-to farmer model, e.g., Esoko.
- Business-to-business-to-consumer model is used between market-led ICT platform information providers (Ignitia), mobile telecommunication networks (MTN Ghana), and farmers (Chudaska 2018). Another example is farmers on the ADVANCE II project who are connected to MTN Ghana which runs the promotion sale of simple mobile phones to enable farmers to receive CIS (Mohammed 2018).
- Direct services to farmers, e.g., Vodafone farmer club where farmers pay US$ 0.20 per month as a direct debit from their call credits to receive CIS. Subscribers also receive free airtime to call members of the group (Partey et al. 2019, 2020).
- Business to nucleus farmer and outgrower schemes is an approach that is becoming common where CIS is provided as an embedded service model. In this approach, agri-input companies, in particular, provide information about farm inputs, and CIS is added as a package. The cost of the CIS is embedded as an unidentified component of the sales price. An example of this governance approach is used by the USAID-funded ADVANCE project to promote the use of input supplies, improved seeds, and CIS (MEAS 2012; Mohammed 2018). Wienco Ghana also provides CIS with this model in the areas of cash and food

crops for both small- and large-scale farmers in the agricultural value chain (Naab et al. 2019).

Informal governance approach for the provision of CIS includes traditional institutional arrangement used by farming communities. These are in the form of rules or norms that the farming community adheres to (Slavova and Karanasios 2018).

Discussion of Emerging Issues

First, informal channels of information delivery still hold prominence among small-holder farmers as a means for interaction. For instance, the provision of CIS by agricultural extension agents, workshops, participatory approaches, and the use of social networks are still prominent (Nyantakyi-Frimpong 2019a). Furthermore, information sources in farming communities are often embodied in knowledgeable people and shared through interactions occurring at the market, religious meetings, bus stops, community centers, and other places of meetings. Most often, the CIS is received by one person in the community, and it is shared with other members of the farming community (Kirbyshire and Wilkinson 2018). This has implications on credibility, trust, and sustainability of financial models of actors who provide CIS through ICTs. Due to the need for interactions among smallholder farmers, prefer-ences for ICTs for the delivery of CIS still hinge on interactive technologies which provide an opportunity for farmers to ask questions and share their experiences. This suggests the need to support innovations on the use of interactive ICTs for the delivery of CIS.

Despite the limited use of new ICTs among farmers, mobile phones, in particular, have helped farmers to communicate CIS to other farmers, thereby enhancing interactions and reinforcing interpersonal relationships and norms of openness among farmers. First, mobile phones strengthened bonds in the rural community by enabling information exchanges between family, farmers, and intermediaries. Second, mobile phones complement informal ways of communicating among farmers. Therefore, norms of openness, inclusiveness, and information-sharing are strengthened rather than challenged. The findings suggest that the informal and formal channels of CIS delivery support smallholder farming institutional logic than ICTs destroying informal institutions.

Additionally, intermediaries who provide CIS with ICTs are able to exploit complementarities among digital tools and informal channels of communication. They are able to fuse characteristics and logics of informal communication channels with ICTs such as call center, IVR, and radios which relate to established patterns of interaction among smallholder farmers. The use of mobile phones, SMS alert, or peep call during radio programs is also an example. The combination of informal and ICTs helps to increase farmers' access to CIS.

Despite the evolution of sophisticated ICTs, the provision of CIS with ICTs is limited to simple ones like mobile phones, radio, and television. Thus, farmers rarely use social media, weather apps, and websites to receive useful CIS. Some farmers are unable to operate the phone without the assistance of relatives. This leaves the potentials of ICTs untapped for the delivery of CIS, or it results in the use of platform intermediaries. These challenges are attributed to increased sophistication of ICTs without being tailored for local farmers with literate and visual challenges and limited training on the use of ICTs (Alemna and Sam 2006; Sarku et al. forthcoming). The unavailability of local content ICTs also affects its usage as technologies are originally coded in English. In addition, many languages in Ghana use characters that are not found on the keyboards of most ICTs. Furthermore, the provision of CIS for the target group requires specific local content which affects the scalability of CIS to other agroecological locations in Ghana. There are issues regarding the affordability of ICTs such as smartphones, expensive call charges, and data (Partey et al. 2020; Etwire et al. 2017). All these challenges have implications on information asymmetry as some farmers can use ICT and access tailored CIS, while others still rely on their local knowledge for forecasting.

Findings in this chapter show that the delivery of CIS with ICT platforms are mostly piloted or funded by donors. At the end of the project, farmers are expected to subscribe to continue to receive information at a fee. However, their willingness to pay for the information raises questions about the continuity of CIS delivery. Many of the projects are designed with a top-down approach, without collaborations from farmers at the implementation stage. This raises questions on upscaling from pilot projects to larger farming communities. A limitation of relying on donors to fund the delivery of CIS with ICTs is that specific agroecological zones are targeted for a short period (McNamara et al. 2014). This affects the number of farmers who are reached within the specified period. Since climate change and variability impacts are experienced differently in specific agroecological zones in Ghana, it will be important to have nationwide CIS tailored for all farmers.

Conclusion

This chapter examined the delivery of CIS with digital tools to smallholder farmers by reviewing existing literature on the sector. Overall, our rapid scoping assessment found very positive results on the use of various digital tools for the collection, processing, and delivery of CIS to farmers. The delivery of CIS with ICTs was used mostly by men and during the farming season; a variety of ICTs are used to deliver CIS due to the level of demand for information. Old technologies, like radio, remain the most relevant and cost-efficient ICT platforms for the delivery of CIS. Some ICTs remain underused by farmers calling for the role of ICT platform intermediaries like agricultural extension agents and field agents. The delivery of CIS with ICTs results in changes in institutional logics in farming communities in the form of new interactions, information exchange, and other forms of innovation intermediation. Overall, informal information delivery channels coexist with ICT modes and are sometimes blended in various ways to provide CIS. The analysis in this chapter provides a coherent overview on the application of digital tools for the delivery of

CIS for smallholder farming in Ghana. The emerging issues identified in the literature have contributed to the identification of areas for future research. These include (1) exploration of appropriate financial models for the sustainability and scalability of CIS delivery with ICTs; (2) application of ICTs for the delivery of CIS for other value chain actors such as processors, transporters, and aggregators; (3) increase research on how ICTs are used for data collection and production at the community level; and (4) assessment of the quality of CIS provided to smallholders.

References

Acquah HG, Onumah EE (2011) Farmers perception and adaptation to climate change: an estimation of willingness to pay. Agris On-line Papers Econ Inform 3(4):31–39

Adiku SG, Debrah-Afanyede E, Greatrex H, Zougmore RB, MacCarthy DS (2017) Weather-index based crop insurance as a social adaptation to climate change and variability in the Upper West Region of Ghana. Working paper no 189. CGIAR Research Program on Climate Change, Agriculture and Food Security, Copenhagen

Adjabui JA (2018) Farmers' willingness to participate and pay for, and agricultural extension officers' disposition to communicate weather index-based insurance scheme in Ghana: the case of the upper east region. Master thesis, Massey University

Adjin-Tettey TD (2013) The perception and usage of weather forecast information by residents of African concrete products (ACP) estates and farmers in selected communities around Pokuase in the Ga west municipality of Ghana. Int J ICT Manage I(3), 139–149.

Agyekumhene C, de Vries JR, van Paassen A, Macnaghten P, Schut M, Bregt A (2018) Digital platforms for smallholder credit access: the mediation of trust for cooperation in maize value chain financing. Wagening Journal of Life Sciences, 86–87, 77–88. https://doi.org/10.1016/j.njas.2018.06.001

Aker JC (2011) Dial "A" for agriculture: a review of information and communication technologies for agricultural extension in developing countries. Agric Econ 42:631–647

Aker PG, Gilligan J, Hidrobo DM, Ledlie N (2018) Paying for digital information: assessing farmers' willingness to pay for a digital agriculture and nutrition service in Ghana Other area references. In: 30th International conference of agricultural economist

Akudbillah NA (2017) Modelling the socio-economic benefits of the adoption of climate information: an innovative approach to subsistence farmer adaptation to climate change in Garu-Tempane District, Ghana. Master's thesis, University of Bergen

Akudugu MA, Guo E, Dadzie SK (2012) Adoption of modern agricultural production technologies by farm households in Ghana: what factors influence their decisions. J Biol Agric Healthc 2(3). Retrieved from https://www.iiste.org/Journals/index.php/JBAH/article/view/1522

Alemna AA, Sam J (2006) Critical issues in information and communication technologies in Ghana. Inform Dev 22(4). https://doi.org/10.1177/0266666906074181

Alhassan RM, Egyir IS, Abakah J (2013) Farm household level impacts of information communication technology (ICT)-based agricultural market information in Ghana. J Dev Agric Econ 30 (5):161–167

Anaglo JN, Boateng SD, Boateng CA (2014) Gender and access to agricultural resources by smallholder farmers in the upper west region of Ghana. J Educ Pract 5(5). http://www.iiste.org

Anuga SW, Gordon C (2016) Adoption of climate-smart weather practices among smallholder food crop farmers in the Techiman municipal: implication for crop yield. Res J Agric Environ Manage 5(9):279–286

Anuga SW, Gordon C, Boon E, Musah-Issah JS (2019) Determinants of climate smart agriculture (CSA) adoption among smallholder food crop farmers in the Techiman municipality, Ghana. J Geogr 11(1):124–139. https://doi.org/10.4314/gjg.v11i1.8

Asenso-Okyere K, Mekonnen DA (2012) The importance of ICTs in the provision of information for improving agricultural productivity and rural incomes in Africa. Working Paper 2012–

015. United Nations Development Programme's Regional Bureau for Africa, Addis Ababa. Available at http://web.undp.org/africa/knowledge/WP-2012-015-okyere-mekonnen-ict-produc tivity.pdf

Asuru S (2017) The new philanthropy and smallholder farmers' livelihoods. A case study of the Alliance for a Green Revolution in Africa (AGRA) in the northern region of Ghana. Doctoral thesis, University of Bradford

Barber J, Mangnus E, Bitzer V (2016) Harnessing ICT for agricultural extension. KIT working paper 4. Amsterdam. Accessed from https://www.kit.nl/wp-content/uploads/2019/10/KIT_ WP2016-4_Harnessing-ICT-for-agricultural-extension.pdf

Bell M (2015) Information and communication technologies for agricultural extension and advisory services. ICT–powering behavior change for a brighter agricultural future. MEAS Project, Urbana

Bennett WL, Segerberg A (2012) The logic of connective action: digital media and the personalization of contentious politics, information. Commun Soc 15:739–768. https://doi.org/10.1080/1369118X.2012.670661

Buytaert W, Zulkafli Z, Grainger S, Acosta L, Alemie TC, Bastiaensen J et al (2014) Citizen science in hydrology and water resources: opportunities for knowledge generation, ecosystem service management, and sustainable development. J Front Earth Sci 2:1–21. https://doi.org/10.3389/feart.2014.00026

Caine A, Dorward P, Clarkson G, Evans N, Canales C, Stern D (2015) Review of mobile applications that involve the use of weather and climate information: their use and potential for Smallholder Farmers. CCAFS working paper no150. CGIAR Research Program on Climate Change, Agriculture and Food Security (CCAFS), Copenhagen. Available online at: http://www.ccafs.cgiar.org

Chapman R, Blench R, Kranjac-Berisavljevic G, Zakariah ABT (2003) Rural radio in agricultural extension: the example of vernacular radio programmes on soil and water conservation in Ghana. Network paper no 127. Overseas Development Institute, London

Chaudhury M, Kristjanson P, Kyagazze F, Naab JB, Neelormi S (2012) Participatory gender-sensitive approaches for addressing key climate change-related research issues: evidence from Bangladesh, Ghana, and Uganda. Working paper 19. CGIAR Research Program on Climate Change, Agriculture and Food Security (CCAFS), Copenhagen

Chudaska R (2018) Informational governance: exploring how the flow of water-related information affects farming practices and decision-making in Ada east. Master thesis, Wageningen University

Cieslik K, Leeuwis C, Dewulf A, Feindt P, Lie R, Werners S et al (2018) Addressing socio – ecological development challenges in the digital age: environmental virtual observatories for connective action. NJAS Wagening J Life Sci 86–87:2–11

Clarksona G, Dorward P, Osbahra H, Torgbor F, Kankam-Boadu I (2019) An investigation of the effects of PICSA on smallholder farmers' decision making and livelihoods when implemented at large scale–the case of northern Ghana. Clim Serv 14:1–14

Dilling L, Lemos MC (2011) Creating usable science: opportunities and constraints for climate knowledge use and their implications for science policy. Glob Environ Chang 21:680–689

Dinku T, Thomson MC, Cousin R, del Corral J, Ceccato P, Hansen J, Connor SJ (2018) Enhancing national climate services (ENACTS) for development in Africa. Clim Dev 10(7):664–672. https://doi.org/10.1080/17565529.2017.1405784

Dougill AJ, Whit S, Stringer LC, Vincent K, Wood BT, Chinseu EL et al (2017) Mainstreaming conservation agriculture in Malawi: knowledge gaps and institutional barriers, 195. https://doi.org/10.1016/j.jenvman.2016.09.076

Drafor I (2016) Access to information for farm-level decision-making. J Agric Food Inf 17(4):230–245. https://doi.org/10.1080/10496505.2016.1213170

Drafor I, Atta-Agyepong K (2005) Local information systems for community development in Ghana. In: Dixon J, Wattenbach H, Bishop-Sambrook C (eds) Improving information flows to the rural community. FAO, Rome, pp 5–19

Dumenu WK, Obeng EA (2016) Climate change and rural communities in Ghana: social vulnerability, impacts, adaptations and policy implications. Environ Sci Policy 55:208–217. https://doi.org/10.1016/j.envsci.2015.10.010

Duncombe R (2016) Mobile phones for agricultural and rural development: a literature review and suggestions for future research. Eur J Dev Res 28:213–235. https://doi.org/10.1057/ejdr.2014.60

Eitzingera A, Cocka J, Atzmanstorferc K, Binder CR, Läderacha P, Bonilla-Findjia O, Bartling M, Mwongeraa C, Zuritae L, Jarvisa A (2019) GeoFarmer: a monitoring and feedback system for agricultural development projects. Comput Electron Agric 158:109–121

Etwire PM, Buah S, Ouédraogo M, Zougmoré R, Partey ST, Martey E, Djibril SD, Bayala J (2017) An assessment of mobile phone-based dissemination of weather and market information in the Upper West Region of Ghana. J Agric Food Secur 6:8. https://doi.org/10.1186/s40066-016-0088-y

Federspiel S (2013) Climate change adaptation planning, implementation and evaluation: needs, resources and lessons for the 2013 National Climate Assessment. University of Michigan School of Natural Resources and Envirnment, Ann Arbor

Gbetibouo G, Hill C, Abazaami J, Mills A, Snyman D, Huyser O (2017) Impact assessment on climate information services for community-based adaptation to climate change: Ghana country report. CARE international adaptation learning program, Ghana. http://careclimatechange.org/wp-

Gotamey A, Shankar A, Toomasson T (2018) Rain forecasting service for rural Ghana implementation prototype report. ICT4D: Information and Communication Technology for Development (X_405101), Amsterdam

Graham R, Visman E, Wade S, Amato R, Bain C, Janes T, Leathes B, Lumbroso D, Cornforth R, Boyd E, Parker D (2015) Scoping, options analysis and design of a 'Climate Information and Services Programme' for Africa (CIASA). Lit Rev. https://doi.org/10.12774/eod_cr.may2015.grahamr

Hansen JW, Vaughan C, Kagabo DM, Dinku T, Carr ER, Körner J, Zougmoré RB (2019) Climate services can support African Farmers' context-specific adaptation needs at scale. Front Sustain Food Syst 3(21). https://doi.org/10.3389/fsufs.2019.00021

Ibrahim N, Teye MK, Alhassan H, Adzawla W, Adjei-Mensah C (2019) Analysis of smallholder farmers' perceptions on climate change, preference and willingness-to-pay for seasonal climate forecasts information in Savelugu Municipality, Ghana. Asian J Environ Ecol 9(1):1–11, Article no.AJEE.47872 ISSN: 2456-690X

IPCC (2014) Climate change 2014: impacts, adaptation, and vulnerability. Part A: global and sectoral aspects. In: Field CB, Barros VR, Dokken DJ, Mach KJ, Mastrandrea MD, Bilir TE, Chatterjee M, Ebi KL, Estrada YO, Genova RC, Girma B, Kissel ES, Levy AN, MacCracken S, Mastrandrea PR, White LL (eds) Contribution of working group II to the fifth assessment report of the Intergovernmental Panel on Climate Change. Cambridge University Press, Cambridge, UK/New York, 1132 pp

Jost C, Kyazze F, Naab J, Neelormi S, Kinyangi J, Zougmore R et al (2016) Understanding gender dimensions of agriculture and climate change in smallholder farming communities. J Clim Dev 8:133–144

Kadi M, Njau LN, Mwikya J, Kamga A (2011) The state of climate information services for agriculture and food security in West African Countries. CCAFS working paper no 4. Copenhagen. Available online at: http://www.ccafs.cgiar.org

Karpouzoglou T, Zulkafli Z, Grainger S, Dewulf A, Buytaert W, Hannah DM (2016) Environmental virtual observatories (EVOs): prospects for knowledge co-creation and resilience in the information age. Curr Opin Environ Sustain 18:40–48. https://doi.org/10.1016/j.cosust.2015.07.015

Kirbyshire A, Wilkinson E (2018) Challenging assumptions: What impact are NGOs having on the wider development of climate services? Research, reports and studies. Overseas Development Institute. https://www.odi.org/sites/odi.org.uk/files/resources-documents/...

Leeuwis C (2004) Communication for rural innovation: rethinking agricultural extension. Blackwell Science, Oxford

Lemos MC, Kirchhoff CJ, Ramprasad V (2012) Narrowing the climate information usability gap. J Clim Change 2(11):789–794

Materia VC, Giarè F, Klerkx L (2015) Increasing knowledge flows between the agricultural research and advisory system in Italy: combining virtual and non-virtual interaction in communities of practice. J Agric Educ Ext 21:203–218. https://doi.org/10.1080/1389224X.2014. 928226

McNamara P, Dale J, Keane J, Ferguson O (2014) Strengthening pluralistic agricultural extension in Ghana: a MEAS Rapid Scoping Mission, Accra. Modernizing extension and advisory services discussion paper. USAID, Illinois.

McOmber C, Panikowski A, McKune S, Bartels W, Russo S (2013) Investigating climate information services through a gendered lens. CCAFS working paper no 42. CGIAR Research Program on Climate Change, Agriculture and Food Security (CCAFS). Copenhagen. Available online at: http://www.ccafs.cgiar.org

Mills A, Huyser O, van den Pol O, Zoeller K, Snyman D, Tye N, McClure A (2016) UNDP market assessment: revenue generating opportunities through tailored weather information products. UNDP, New York. License: Creative Commons Attribution CC BY 3.0 IGO

Modernizing Extension and Advisory Services (MEAS) (2012) Rapid appraisal of the ICT for agricultural extension landscape in Ghana. Retrieved from the MEAS website: http://www. meas-extension.org

Mohammed HN (2018) Digital tools and smallholder agriculture: the role of ICT-enabled extension services in rural smallholder farming in northern Ghana. Master of Arts in Development Studies, International Institute of Social Studies

Mol APJ (2008) Environmental reform in the information age – the contours of informational governance. Cambridge University Press, Cambridge

Munthali N, Leeuwis C, van Paassen A, Lie R, Asare R, van Lammeren R, Schut M (2018) Innovation intermediation in a digital age: comparing public and private new-ICT platforms for agricultural extension in Ghana. Wagening J Life Sci 86–87:64–76

Naab FZ, Abubakari Z, Ahmed A (2019) The role of climate services in agricultural productivity in Ghana: the perspectives of farmers and institutions. Clim Serv 13:24–32

Nantui F, Nketiah MP, Darko D (2014) Farmers' willingness to pay for weather forecast information in Savelugu-Nanton municipality of the northern region. RJOAS 12(36):34

Nyadzi E (2020) Best of both worlds: co-producing climate services that integrate scientific and indigenous weather and seasonal climate forecast for water management and food production in Ghana. PhD thesis, Wageningen University

Nyamekye AB (2020) Towards a new generation of climate information systems information systems and actionable knowledge creation for adaptive decision-making in Rice Farming Systems in Ghana. PhD thesis, Wageningen University

Nyantakyi-Frimpong H (2013) Indigenous knowledge and climate adaptation policy in northern Ghana. The Africa portal backgrounder, number 48. Center for International Governance Innovation, Waterloo

Nyantakyi-Frimpong H (2019a) Unmasking difference: intersectionality and smallholder farmers' vulnerability to climate extremes in Northern Ghana. Gender Place Cult. https://doi.org/10. 1080/0966369X.2019.1693344

Nyantakyi-Frimpong H (2019b) Combining feminist political ecology and participatory diagramming to study climate information service delivery and knowledge flows among smallholder farmers in northern Ghana. Appl Geogr 112:102079

Nyarko Y, Hildebrandt N, Romagnoli G, Soldani E (2013) Market information systems for rural farmers evaluation of Esoko MIS – year 1 results. Blog Post, Cent Technol Econ Dev NYU. http://www.nyucted.org/archives/1108

Omoine H, Chen HM, Akhtar I (2013) RainCast: a voice based weather service for Rural Ghana. Accessed from https://w4ra.org/wp-content/uploads/2018/08/Raincast.pdf

Padgham J, Devisscher T, Togtokh C, Mtilatila L, Kaimila E, Mansingh I, Agyemang-Yeboah F, Obeng FK (2013) Building shared understanding and capacity for action: insights on climate risk communication from India, Ghana, Malawi, and Mongolia. Int J Commun 7:970–983. 1932–8036/20130005. Available at http://ijoc.org

Partey ST, Nikoi GK, Ouédraogo M, Zougmoré RB (2019) Scaling up climate information services through public-private partnership business models: an example from northern Ghana. CCAFS Info note. https://hdl.handle.net/10568/101133

Partey ST, Dakorah AD, Zougmoré RB, Ouédraogo M, Nyasimi M, Nikoi GK, Huyer S (2020) Gender and climate risk management: evidence of climate information use in Ghana. Clim Chang. https://doi.org/10.1007/s10584-018-2239-6

Perkins K, Huggins-Rao S, Hansen J, van Mossel J, Weighton L, Lynagh S (2015) Interactive radio's promising role in climate information services: Farm Radio International concept paper. CCAFS working paper no 156. CGIAR Research Program on Climate Change, Agriculture and Food Security (CCAFS), Copenhagen. Available online at: http://www.ccafs.cgiar.org

Quaye W, Asafu-Adjaye NY, Yeboah A, Osei C, Agbedanu EE (2017) Appraisal of the agro-tech smart extension model in Ghana, payment options and challenges in ICT-enabled extension services delivery. Int J Agric Educ Ext 3(2):72–84

Sam J, Dzandu L (2016) The use of radio to disseminate agricultural information to farmers: the Ghana agricultural information network system (GAINS) experience. Agric Inf Worldw 11:17–23

Sarku R, Dewulf A, Slobbe VE, Termeer K, Kranjac-Berisavljevic G (2020) Adaptive decision-making under conditions of uncertainty: the case of farming in the Volta delta, Ghana. J Integr Environ Sci 17(1):1–33. https://doi.org/10.1080/1943815X.2020.1729207

Sarku R, Gbangou T, Dewulf A, Slobbe VE (forthcoming) Beyond 'experts knowledge': locals and experts in a joint production of weather App and weather information for farming in the Volta Delta, Ghana. In: Handbook of climate change management: research, leadership, transformation. Springer, New Delhi

Singh C, Urquhart P, Kituyi E (2016) From pilots to systems: barriers and enablers to scaling up the use of climate information services in smallholder farming communities. CARIAA working paper no 3. International Development Research Centre/UK Aid, Ottawa/London. Available online at: http://www.idrc.ca/cariaa

Singh C, Daron J, Bazaz A, Ziervogel G, Spear D, Krishnaswamy J, Zaroug M, Kituyi E (2018) The utility of weather and climate information for adaptation decision-making: current uses and future prospects in Africa and India. Clim Dev 10(5):389–405. https://doi.org/10.1080/17565529.2017.1318744

Slavova M, Karanasios S (2018) When institutional logics meet information and communication technologies: examining hybrid information practices in Ghana's agriculture. J Assoc Inf Syst 19(9):775–812. https://doi.org/10.17705/1jais.00509

Soma K, MacDonald BH, Termeer CJAM, Opdam P (2016) Introduction article: informational governance and environmental sustainability. J Curr Opin Environ Sustain 18:131–139. https://doi.org/10.1016/j.cosust.2015.09.005

Stigter K, Winarto YT, Ofori E, Zuma-Netshiukhwi G, Nanja D, Walker S (2013) Extension agrometeorology as the answer to stakeholder realities: response farming and the consequences of climate change. J Atmos 4:237–253. https://doi.org/10.3390/atmos4030237

Suchiradipta B, Saravanan R (2016) Social media: shaping the future of agricultural extension and advisory services. GFRAS interest group on ICT4RAS discussion paper. GFRAS, Lindau. Accessed from Social media: Shaping the future of agricultural ... – GFRAS. http://www.g-fras.org/knowledge/gfras-publications

Sulaiman RV, Hall A, Kalaivani NJ, Dorai K, Reddy TSV (2012) Necessary, but not sufficient: critiquing the role of information and communication technology in putting knowledge into use. J Agric Educ Ext 18:331–346. https://doi.org/10.1080/1389224X.2012.691782

Tall A, Hansen J, Jay A, Campbell B, Kinyangi J, Aggarwal PK, Zougmoré R (2014) Scaling up climate services for farmers: mission possible. Learning from good practice in Africa and South Asia. CCAFS report no 13. CGIAR Research Program on Climate Change, Agriculture and Food Security (CCAFS), Copenhagen

Usher J, Phiri C, Linacre N, O'Sullivan R, Qadir U (2018) Climate information Services market assessment and business model review. USAID-supported Assessing Sustainability and Effectiveness of Climate Information Services in Africa Project, Washington, DC

Vaughan C, Dessai S (2014) Climate services for society: origins, institutional arrangements, and design elements for an evaluation framework. J Clim Change 5(5):587–603. https://doi.org/10.1002/wcc.290

Vogel C, Steynor A, Manyuchi A (2019) Climate services in Africa: re-imagining an inclusive, robust and sustainable. Clim Serv 15:100107

World Meteorological Organization (2015) Valuing weather and climate: economic assessment of meteorological and hydrological services. WMO-No 1153. WMO, Geneva

Zougmoré RB, Partey ST, Ouédraogo M, Torquebiau E, Campbell BM (2018) Facing climate variability in sub-Saharan Africa: analysis of climate-smart agriculture opportunities to manage climate-related risks. Cahiers Agric 27(3):1–9

6

Community Adaptation to Climate Change: Case of Gumuz People, Metekel Zone, Northwest Ethiopia

Abbebe Marra Wagino and Teshale W. Amanuel

Contents

Abstract

The effect of climate change on agricultural-dependent communities is immense. Ethiopia in which more than 85% of its population is agrarian is affected by climate change. Communities in different parts of the country perceived climate change and practice different climate change adaptation strategies. This chapter was initiated to identify adaptation strategy to the impact of changing

A. M. Wagino (✉)
Mendel University in Brno Project in Ethiopia, Addis Ababa, Ethiopia

T. W. Amanuel
Wondo Genet College of Forestry and Natural Resource, Hawassa University, Hawassa, Ethiopia

climate. Data on a total of 180 households were gathered using structured and semi-structured questioners. Focus group discussion and key informant interview were also used for data collection. Climatic data from the nearest meteorological stations of the area were collected and used in this chapter. The collected data were analyzed using descriptive and inferential statistical methods. The upshot indicated that all the respondent communities experienced at least one of autonomous/self-adaptation strategies to cope and live with the impacts of changing climate. Though 33.6% complained on its accessibility and pricing, 66.4% of the respondents reviled as they do not have any awareness on improved agricultural technologies. The major adaptation strategies identified were collecting and using of edible wild plants and other forest products, hunting, renting/selling of own farm lands, livestock sell, selling of household materials/ assets, collecting and selling of wood and wood products and depending on well-off relatives, using drought-resistant crop variety, changing cropping calendar, replanting/sowing, and increasing farmland size. Nevertheless, the communities are not yet fully aware and accessed to policy-driven options for climate change adaptation. Although they used different autonomous adaptation mechanisms, the households are not resilient to the current and perceived climate change. Finally, based on the findings, the recommendation is that besides encouraging the existing community-based adaptation strategies planned adaptation strategies have to be implemented: such as early-warning and preparedness programs have to be effectively implemented in the area, introduction of different drought-resistant locally adapted food crop varieties, and expansion of large-scale investment in the area has to be checked, and give due recognition to forest ecosystem–based adaptation mechanisms of the local community in the area.

Keywords

Climate change · Adaptation strategy · Autonomous adaptation · Planned adaptations · Forest ecosystem · based adaptation options

Introduction of the Chapter

The world's climate has already changed and will change dramatically. Under the no emission scenario, the average global surface temperature is predicted to increase by 2.8 °C during this century (IPCC 2007). It is now a challenge for the entire world with growth, poverty, food security, and stability implications. The demand of the hour is to adapt to the changing climate and work together to find mitigation options so that no further damage is done (IPCC 2007).

However, the impacts of existing and predicted changes in climate vary across economies. Poor countries can incur huge costs from a small deviation in climate, particularly due to their dependency on climate-sensitive sectors (Agriculture), poor adaptive capacity, lack of necessary technology, and lack of resources to deal with the stress (Seo and Mendelsohn 2008). For instance, because of significant dependence on the agricultural sector for production, employment, and export revenues,

Ethiopia is seriously threatened by climate change, which contributes to frequent drought, flooding, and rising average temperatures (Emerta 2013). Severe droughts often are followed by severe food insecurity, population dislocation, family separation, and erosion of the sociocultural fabric (Berhe and Butera 2012).

Like most parts of Ethiopia, climate change brings newer and more complicated challenges to people in Benishangul Gumuz region, having devastating implications for the peoples of the region in general and the Gumuz communities in particular (Emerta 2013). The incidence of crop pests and disease increased from time to time, the size of their forest ecosystem is threatened due to farmland and settlement encroachment by other ethnic groups (Abbute 2002). The existing socioecological system is reduced to support the Gumuz community to cope and live with the new climate stimuli as usual by traditional means. This coupled with their temporally and spatially cyclical agricultural system that involves clearing of land – usually with the assistance of fire – followed by phases of cultivation and fallow periods with the help of rudimentary labor-intensive farm tools threatened the lives of the people in the region. Particularly, the study area is highly affected by climate change and variability. As Sani et al. (2017) indicated the overall natural resources base of the region is highly degraded. Thus, people in the region are facing a variety of shocks and become vulnerable. However, the Gumuz community in the area has been responding to climate change through various adaptation strategies. However, there was no scientific study that substantiates or supports the existing adaptation strategies practiced by the Gumuz people in the area.

Identification of traditional risk mitigation and coping strategies that explicitly show elasticity of the community to existing disturbances in order to evaluate the suitability of current adaptive behaviors, as well as assess the adaptation deficit of local communities in view of increasing climate variability is very important. It would help in identifying those available adaptive measures/options that need to be built on and strengthened, as well as innovative adaptation strategies that add value to current climate risk mitigation and coping behaviors, by effectively addressing adaptation constraints experienced by communities (Berhe and Butera 2012). The knowledge of local adaptation strategies are essential to cope, adapt, and live with current and perceived extreme climate variabilities/changes through building their indigenous resilience mechanisms (WFP and FAO 2012). Therefore, knowing and building community adaptation practice is indispensable to live in the midst of the change. In addition, despite the huge potential that traditional knowledge offers for climate change adaptation, research efforts on the effectiveness and appropriateness of this knowledge have, so far, been limited. Furthermore, there are no proven approaches to integrate traditional coping mechanisms into mainstream development efforts (Berhe and Butera 2012).

Berhe and Butera (2012) indicated that drought and climate variability are part of the natural cycle in lowland Ethiopia, and the communities do have an array of traditional coping mechanisms. Indigenous peoples are excellent observers and interpreters of change on the land, sea, and sky. Moreover, indigenous knowledge provides a crucial foundation for community-based adaptation and mitigation actions that can sustain resilience of social–ecological systems at the interdependent local, regional, and global scales. However, the ability of a community to maintain a certain level of well-being in the face of risks depends on the resource options available to that community/household to make a living and on their ability to handle risks (Alinovi et al. 2009). Therefore, this study aimed at investigating the climate change adaptation strategies practiced by the Gumuz people in response to its adverse effects and analyzing determinants of the use of adaptation strategies in copping, adapting, and living with observed and perceived climate variability in the area.

Description

Location: Metekel Zone is located in the Benishangul Gumuz National Regional State (BGNRS) (Fig. 1). The zone occupies an estimated total area of 22,028 km² . Geographically it is located between 09.17° and 12.06° north latitude and 34.10° and 37.04° east longitude. The zone encloses seven woredas/districts, namely Bullen, Dibate, Dangur, Guba, Mandura, Pawi, and Wembera. The Addis Ababa-Guba and Chagini to Wombera all-weather road provide the primary access to the area. In the present administrative context, most of the Gumuz inhabit Metekel zone to the north and Kamashi zone to the south of the Abbay/Nile River.

Agro-ecologically, the zone is mostly classified as 82% lowland (kola), 10% midland (woina-dega), and 8% highland (dega) with an average rainfall of 1,275 mm

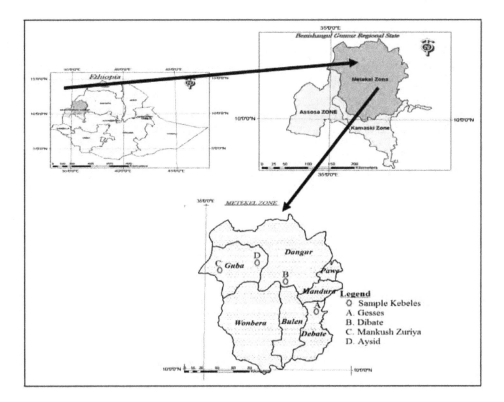

Fig. 1 Map of Metekel zone

per annum and an altitude range of 500–2731 meter above sea level. The Gumuz, who constitute the most numerous ethnic group of the area, mostly inhabit the lowlands in all seven districts. The total population of the zone was 276,367 (male 139,119 and female 137,248) of which 238,752 are rural setup while the remaining 37,615 are urban dwellers (BGRSDGA 2010). The land use pattern is estimated as 79% forestland, 7% cultivated land, 7% cultivable land, and 7% nonutilizable land.

The annual average temperature ranges from 16.2 °C to 32.5 °C with annual mean rainfall of 1,607.8 mm, where the annual rainfall months ranging from May to October (EARO-PRC 2000). The zone has a unimodal rainfall pattern, with an extended rainy season, from March to September. The peak rainy season is from July to August. The coldest months are December and January whereas March and April are the hottest months of the area (Esayas 2003). The mean monthly available meteorological data obtained from Ethiopian Metrological Agency for the four stations namely Bullen, Mandura, Debate, and Guba for the period of 1972 to 2013 is presented in Fig. 2.

Vegetation: Generally, about 55% of the total land area of the region is covered with different vegetation and forests. Bamboo, incense, and gum trees are the major forest types. Forests are important sources of construction material, fuel wood, and food, particularly for the indigenous communities (Benishangul Gumuz Region Food Security Strategy 2004). The original plant cover of the Metekel zone consti-tutes dense hyparrenia, dense bamboo thickets, and scattered trees and of arboreal and thick shrub by formations along the water ways covering areas of various sizes (Dieci and Viezzoli 1992). Degradation of forest resources is increasing at an alarming rate due to various limitations. Encroachment, forest fires, absence of well-defined land use policy, and intensive resettlement programs that took place during the past government regime are some of the main causes for the depletion of natural resources in the region.

Socioeconomic condition: The regional economy depends on agriculture which accounts for 93.2% of the economically active population. Shifting cultivation is the major economic activity of the Gumuz community. Shifting cultivation is broadly

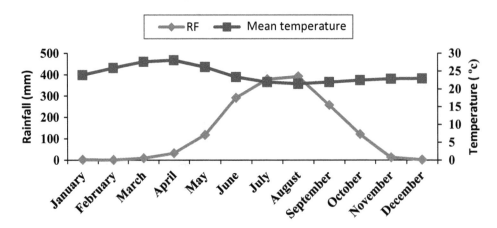

Fig. 2 Mean monthly rainfall and temperature of the study area

defined as "any temporally and spatially cyclical agricultural system that involves clearing of land usually with the assistance of fire followed by phases of cultivation and fallow periods. The subsidiary livelihood sources include livestock raising, gathering wild foods, fishing, honey production and collection, traditional gold mining, hunting, handicrafts, petty trade and charcoaling" (Anonymous 2004, 2011).

Rudimentary labor-intensive farm tools usage, prevalence of crop diseases, pests and weeds, declining soil fertility, inadequate use of improved inputs, erratic rainfall, human diseases such as malaria, poor rural infrastructure facilities like market and road, absence of credit services, and poor working culture of the communities largely due to use of labor-intensive farm tools and low awareness resulted very limited crop production and productivity in the area.

Information Assortment Techniques

A multistage sampling technique was used to collect information/data. In the first stage, out of the seven woredas (districts) where the Gumuz ethnic groups predominantly live in the zone, three woredas were purposively selected to include different attributes. The Gumuz community lives almost in the same agro-ecological zones in all woredas. Then within the zone Debate, Dangur, and Guba woredas were selected for this study. In the second stage of sampling, kebeles within the selected woredas were selected according to settlement patterns, farming practices, crop varieties, socioeconomic aspects, climate problems, and disasters besides biophysical features.

Accordingly, Gessess kebele consisted of 260 households from Debate woreda, Debate kebele consisted of 228 households from Dangur woreda, Aysid consisted of 244 households, and Mankush-zuriya kebeles consisted of 107 households from Guba woreda were selected. The numbers of farmers in each sample peasant association were different, specific numbers of respondents were selected with probability proportionate to size (PPS) random sampling technique to ensure representativeness of the population. Consequently, more than 20% sampling intensity was used from each kebele, accounting a total of 180 sample households and interviewed.

Data collection: Both primary and secondary sets of data were collected. The primary data were collected by using tools of household survey (using structured questionnaire and in-person interview), focus group discussion (FGD), and key informant interview (KII). The valuable secondary data were also obtained from various sources including previous scientific studies and reports from Zonal level agricultural department and other concerned organizations. In addition, the climate data (rainfall and temperature data) was obtained from the National Metrology Agency for the year 1972–2013.

Examination: SPSS statistical software (IBM-SPSS software; version 20) was used throughout the statistical analysis. Both qualitative and quantitative techniques were used to inquiry and presented in the forms of interpretations, comparisons, and arguments. In addition, the quantified information collected using PRA (Participatory Rural Appraisal) tools also presented in figures and percentage. The quantitative analyses made use of all descriptive, correlation, and inferential statistical techniques. The descriptive statistical techniques applied in the study include percentages and graphs. Correlation statistics was used to determine the associations of the major autonomous adaptation strategies, major determinants of community resilience in copping, adopting and living with the impacts of current and perceived climate extreme events, and the interaction of extracted community resilience building blocks presented in the form of tables.

In all cases the significance of the result was also tested using statistical T-test and chi-square Pearson test in determining the p-value to define the significant variation of the obtained results among sample woredas and relative wealth categories of respondent households.

Adaptation Strategy

Crop-Based Adaptation Strategies

Almost all (98.9%) of the respondent households practice rain-fed agriculture. Out of 180 respondents only two of them, 1.1%, reported to have owned an irrigated land of 0.5 hactar each. The major crop-based autonomous adaptation options include: (1) extension of fallowing period, (2) delaying of sowing period and/or sowing two to three times on a specific plots of land for a single harvest following availability of moisture, (3) adoption of hardy and early maturing crops varieties, and (4) increase farm land size (Table 6).

About 75% of the respondents use fallowing crop production system of farming practice (Table 1). Plots are cultivated for 1 or 2 years, then left to recover to woodland by lying fallow for 5 or more years. It is a system which involves clearing of irregular plots at some distance from the village by cutting and burning off trees and shrubs prior to tilling with the hoe. The FDG and KII participants indicated as farmers extend their fallowing period as an option to the effect of climate variability. They perceived that extending fallowing period will increase moisture holding capacity and soil fertility of the land which further increase productivity of the land through reducing weed and pest/disease infestation. The contribution of fallowing in reducing crop weed and pest/disease was mainly reported in Dangur woreda. It is also indicated by Nyong et al. (2007) as natural mulches moderate soil temperatures and extremes, suppress diseases and harmful pests, and conserve soil moisture. However, the practice may increase emission of GHG (Green House Gases) since it involves clearing and burning of forest covers. Though the advent of chemical fertilizers usage is highly encouraged by local level government, local farmers largely depended on organic farming with zero tillage practice which is also capable of reducing GHG emissions.

The FGD participants in GubaWoreda of both Mankush-zuriya and Aysid kebeles testified that almost 100% of the farmers changed the variety of their provenance local crop seeds as an option of adaptation to climate change to lead their subsistence life. As a result of increasing trends of temperature which is subjected to increase the drying period of the season and declining effects of rainfall amount, they forced to change the variety of their provenance crop seeds, locally named as "Bobbe/ yeshenkuit" into "Tirkuwash," a sudan crop variety (both crops belong to sorghum bicolour crop type), because of its adaptation to drought and having short period of time on field for harvesting than the local variety. However, they also explained as they preferred the local variety of their own because of its test difference and ease of management on field. During the discussion, it was understood that the local verities were not easily managed to get and visualized. Besides all the respondent households and FGD participants indicated to sow crop seeds at least two to three times per a single season on a specific plots of lands due to irregular rainfall in the area.

Usage of diversified crop production system through mixed cropping (intercropping) is also mentioned by FGD participants and by 66.6% of respondent farmers (Table 1). This also includes planting more drought-resistant crops with traditionally adopted early-maturing varieties, for example, the farmers sow pumpkin with other lowland crops and start using of its leaf early at the time when it starts emerging leaf. Furthermore, 93.3% of the respondents have a plan to increase their farm land. The factor that drives respondent households to increase their farm land was mainly due to climate variability that induced food insecurity–related issues (Table 2). As indicated in Table 2, 69.9% of the respondent indicated shortage of

Table 1 Agricultural crop production system

| | Yes | | No | |
Production system	Count	%	Count	%
Intercropping production system	120	66.6	60	33.4
Crop rotation production system	108	60	72	40
Fallowing production system	135	75	45	25

Table 2 Factors that drive households to increase their farmlands

| Description | | | Factors | | | | |
			Low production	Low market prices	Food insecurity	Other (specify)	Total
Woreda	Dibate	Count	36	0	16	4	56
		% within Woreda	64.3	0	28.6	7.1	100
	Dangur	Count	57	0	2	0	59
		% within Woreda	96.6	0	3.4	0	100
	Guba	Count	30	2	29	0	61
		% within Woreda	49.2	3.3	47.5	0	100
Total		Count	123	2	47	4	176
		% within Woreda	69.9	1.1	26.7	2.3	100

crop production both for market and household consumption, while 26.7% depicted the existence of food insecurity in general. The rest 2.3% of the households also reported as others (to support their dependents, ancestors).

To this end, as it was reported by one of key informant interviewee at Dangurworeda, and later confirmed by development agents of the surveyed kebeles, during the critical food shortage months the Gumuz farmers have been taking "one quintals of maize to return two quintals of sesame (locally named as Selit), the known cash crops of the area, aftershocks within the same harvesting year," unless he/she is forced to give a reasonable land of his/her own to the guy so that he/she can plow until return the amount of sesame as promised. This is obvious that it affects the capacity of the community to adapt to the current impacts of climate variability or changes, though they use the system to safe their life when the family face a food shortage induced by variation of rainfall.

Although all of the respondents are engaged in rain-fed agricultural practice, a limited effort made as a planned adaptation option. In this regards, Table 3 shows that 63.8% of the respondents do not used any kinds of agricultural inputs to maximize their production from agriculture sector, while the rest used either of agricultural inputs indicated in Table 3. As it was reported by KII the overall improved agricultural practice by the Gumuz people in the area are still at demonstration stage.

Table 4 also showed that almost more than 67.5% of the surveyed households did not have both access and potential to use agricultural inputs with no significance

Table 3 Improved agricultural inputs utilized/practiced or not (in %)

Improved agricultural inputs	Debate		Dangur		Guba		Total	
	Yes	No	Yes	No	Yes	No	Yes	No
Inorganic fertilizer	7	93	10.2	89.8	6.7	93.3	8	92
Improved seeds	7	93	3.4	96.6	35	65	15.1	84.9
Organic fertilizer	10.5	89.5	5.1	94.9	23.3	76.7	13	87
Herbicides chemicals		100	25.4	74.6	1.7	98.3	9	91
Do not used any forms of agricultural inputs	83.9	16.1	57.6	42.4	50	50	63.8	36.2

Table 4 Respondents having access and potential to use agricultural inputs or not

Surveyed woredas		Yes	No	Total
Dibate	Count	11	43	54
	%	20.4	79.60	100
Dangur	Count	23	35	58
	%	39.7	60.3	100
Guba	Count	21	36	57
	%	36.80	63.2	100
Total	Count	55	114	169
	%	32.5	67.5	100

difference among surveyed woredas and/or kebeles, beside low level of the community awareness and skills on application of different improved agricultural technologies as depicted by FGD participants. The rest 32.5% appreciated their own autonomous means adaptation options. Among those who use improved agricultural inputs 36.2% are also reported that they use the inputs only for cash crop production not yet used for food crop production purpose because of its price. Also they are satisfied in using the existing natural soil fertility in the area with no significance difference among sample woredas and respondent wealth category, as reported. Hence, revisiting of promotional approaches/system in the way that the Gumuz community awareness level and capacity developed to adopt and use introduced technologies as a planned adaptation option to boost the return from crop production in the area which further builds the resilience of the community to adapt to the impact of current and perceived climate extremes.

Livestock-Based Adaptation Strategies

In addition to agricultural crop farm, 84.1% of the respondent households have at least one type of animal/livestock husbandry including chicken, with no significant difference among study woredas of sampled kebeles. The household survey and FGD verified as the farmers in the surveyed areas use a mixed farming system besides their dependency on the forest ecosystem to sustain their livelihood. Anonymous (2004) also indicated livestock plays a big role in the livelihood of the community in the areas.

However, 50.3% of respondents indicated the declining trends of livestock production since the last 5 years which is attributed to shortage of feed (73.6%) and stealing of animals (24.7%). Also 35.8% of feed shortage perception was associated with shortage of rainfall. In this regards, 97.2% of respondents use communal grazing land feeding system while 2.8% feed through hay/silage preparation/zero grazing by their own traditional means. FGD and KII indicated that except Debate woreda where stealing of animals was worthily mentioned and looks after their cattle, the rest sends their cattle to areas where permanent water flows exists, just like wild animals. The cattle return to the village at the beginning of June where new grass started to sprout out. Consumption of milk and milk products including eggs are not common for Gumuz people. However, almost all respondents reported they use selective keeping of livestock like goat and donkey since they have a capacity to adopt climate extremes through feeding leafs of bushes and other by-products of prepared local drinks, respectively. The discussion participants reported they use traditional medicine for their animal health care during their visit periodically.

As a means of adaptation options 91% of respondents indicated that they sell animals to adopt with the current climate change–induced stress Table 6. The correlation test results in Table 5 indicated selling of animals negatively correlated with hunting of wild animals and renting/selling of his/her own farm lands. This indicates if a household sells their animals their dependency participation/engagement in hunting of wild animals and renting/selling of his/her

Table 5 Pearson correlations of autonomous adaptation strategies/parameters estimate

Correlations		Collecting edible wild plants	Collecting/ hunting wild animals	Getting support from better-off relatives/freely	Renting/ selling of his/her farm lands	Selling of animals	Selling of available household materials and/or equipments	Collecting and selling different wood and wood products
Collecting edible wild plants	Pearson correlation	1	0.414[a]	−0.029	0.236[a]	0.085	0.071	−0.086
	Sig.		0.000	0.353	0.001	0.131	0.176	0.127
	N	177	177	177	177	177	176	177
Collecting/hunting wild animals	Pearson correlation	0.414[a]	1	0.069	0.377[a]	−0.055	0.035	−0.045
	Sig.	0.000		0.179	0.000	0.235	0.321	0.277
	N	177	177	177	177	177	176	177
Getting support from better-off relatives/freely	Pearson correlation	−0.029	0.069	1	0.115	0.107	0.333[a]	0.234[a]
	Sig.	0.353	0.179		0.064	0.078	0.000	0.001
	N	177	177	177	177	177	176	177
Renting/selling of his/her farm lands	Pearson correlation	0.236[a]	0.377[a]	0.115	1	−0.030	0.089	0.011
	Sig.	0.001	0.000	0.064		0.344	0.120	0.444
	N	177	177	177	177	177	176	177
Selling of animals	Pearson correlation	0.085	−0.055	0.107	−0.030	1	0.234[a]	0.281[a]
	Sig.	0.131	0.235	0.078	0.344		0.001	0.000
	N	177	177	177	177	177	176	177
Selling of available household materials and/or equipments	Pearson correlation	0.071	0.035	0.333[a]	0.089	0.234[a]	1	0.307[a]
	Sig.	0.176	0.321	0.000	0.120	0.001		0.000
	N	176	176	176	176	176	176	176

(continued)

Table 5 (continued)

Correlations		Collecting edible wild plants	Collecting/ hunting wild animals	Getting support from better-off relatives/freely	Renting/ selling of his/her farm lands	Selling of animals	Selling of available household materials and/or equipments	Collecting and selling different wood and wood products
Collecting and selling different wood and wood products	Pearson correlation	−0.086	−0.045	0.234[a]	0.011	0.281[a]	0.307[a]	1
	Sig.	0.127	0.277	0.001	0.444	0.000	0.000	
	N	177	177	177	177	177	176	177

[a]Correlation is significant at the 0.01 level (1-tailed)

Table 6 Autonomous adaptation strategies used by the surveyed households

S. no.	List of adaptation options used	Dibate		Dangur		Guba		Total		P-value	
		No	Yes	No	Yes	No	Yes	No	Yes	Among woredas	Among wealth category
1	Collecting edible wild plants and nontimber forest products	0	32	0	33.3	2.3	33.8	2.3	97.7		
2	Collecting/hunting wild animals	2.3	29	0.6	32.8	9	26	11.9	88.1		
3	Getting support from better-off relatives/freely	8.5	26	18.1	18.3	7.3	30.1	33.9	74.4	P<0.000	
4	Borrow food, or rely on help from a relative/to be back in labor, in cash/in kind after shock	9	23	23.7	9.6	5.6	29.4	38.4	66.6	P<0.000	
5	Renting/selling of his/her farm lands	10.7	21	5.6	27.7	13	22	29.4	70.6		
6	Migrating for seasonal labor selling	17.1	14	29.7	4	15	19.4	62.3	37.7	P<0.000	P<0.008
7	Migrating to other area	19.2	12	32.2	1.1	28	7.3	79.1	20.9	P<0.000	
8	Selling of animals	2.3	29	6.2	27.1	0.6	34.5	9	91	P<0.004	
9	Selling of available household materials and/or equipments	8.5	23	15.3	17.6	5.1	30.1	29	71	P<0.000	
10	Charcoal production and selling	3.4	28	30.1	2.8	23	11.9	56.8	43.2	P<0.000	
11	Fuel wood collection and selling	4	27	26.7	6.8	4	31.2	34.7	65.3		
12	Collecting and selling different wood and wood products	4.5	27	18.6	14.7	1.1	33.9	24.3	75.7	P<0.000	P<0.005
13	Rely on less preferred or less expensive food	12.4	19	15.8	17.5	1.1	33.9	29.4	70.6	P<0.000	
14	Purchase food on credit	28.8	2.8	27.7	5.6	19	16.4	75.1	24.9	p<0.000	P<0.000
15	Consume seed stock that will be needed for next season	15.8	16	18.6	14.7	2.8	32.2	37.3	62.7	p<0.000	
16	Gather wild foods, gather "famine foods," or harvest immature crops	7.9	24	6.8	26.6	7.9	27.1	22.6	77.4		
17	Send household members to eat elsewhere	20.3	11	29.4	4	15	20.3	64.4	35.6	P<0.000	
18	Send household members to beg	27.7	4	31.6	1.7	27	8.5	85.9	14.1		

(continued)

Table 6 (continued)

S. no.	List of adaptation options used	Dibate		Dangur		Guba		Total		P-value	
		No	Yes	No	Yes	No	Yes	No	Yes	Among woredas	Among wealth category
19	Limit portion size at mealtimes	7.3	24	22	11.3	0.6	34.5	29.9	70.1	$p<0.000$	$p<0.000$
20	Restrict consumption by adults in order for small children to eat	12.4	19	25.4	7.9	0.6	34.5	38.4	61.6	$P<0.000$	$p<0.008$
21	Reduce number of meals eaten in a day	6.2	25	18.8	14.8	0.6	34.7	25.6	74.4	$P<0.000$	$P<0.000$
22	Skip entire days without eating	28.8	2.8	22	11.3	22	13	72.9	27.1	$P<0.002$	$P<0.002$
23	Other specifies	23.5	0	58.8	5.9	0	11.8	82.4	17.6		

farm land decrease and vice versa. On the other hand, selling of animals positively correlated with other autonomous adaptation option indicated in Table 5 where it is associated with selling of available household materials or equipments and collecting and selling of different wood and wood products are significant at the 0.01 level (Table 5). This depicted that livestock productions are playing a crucial role in copping the impact of climate change and building the resilience of the farmers in the area.

Forest Resources-Based Adaptation Strategies

The main autonomous coping responses of the household to food shortage caused by drought, erratic rain, and others are presented in Table 6. Most of the respondents (64.8%) indicated that climate stress/shock in the study area resulted in declining of crop and animal production. About 53.5% of the respondents revealed that decline of crop production was associated with variation of rainfall distribution and amount beside other factors. At the time of stress, due to climate change effects, the respondents stated that they are engaged in hunting (85.1%), collecting wild edible plants, and other forest products (88.1%).

On top of this Table 6 it specified that 65.3%, 43.2%, and 75.7% of respondents also engaged on fuel wood selling, charcoal making, and collecting and selling of wood and wood products, respectively. Except collecting and selling of fuel woods which was the same among all woredas and respondents wealth category, the rest of the two (charcoal making and selling of different wood and wood products) have a significant difference among surveyed woredas and wealth category of the respondents. Charcoal making and selling was reported only in Debate woreda. Meanwhile, the large number of respondents from Guba and Debate woreda also conveyed that they engaged on selling of different wood and wood products. In addition, the Pearson correlation test of the major autonomous adaptation strategies indicated in Table 5 shows the existence of strong negative correlation among selling of different wood and wood products with collection of edible wild plants and other forest products, and hunting of wild animals. This may be because of reduced forest resources both in coverage and in its composition to collect edible wild foods and to conduct hunting in Debate and Guba woredas which may be attributed to the existence of large-scale agricultural investment and population pressures than Dangur woredas. As a result, it is also estimated by FGD participants in Debate woreda that the existing tree species reduced by about 40% as an expense of the lost Bamboo species. Moreover, 16.5% of respondents depicted that they compensate any low local market prices, if any, by increasing exploitation of nontimber and timber forest products (NTFP and TFP) besides expanding their agriculture farm land.

Hence, forests are playing a pivotal role in climate change adaptation through building their adaptive capacity which further contributes towards building the resilience of the community in the study area beside its climate change mitigation role. Therefore, provision of due attention in this sector is indispensable to adapt to the current and perceived impacts of changing climate in a sustainable way.

Off-Farm Activities as an Adaptation and Other Copping Strategies

Besides crop, livestock, and forest resource–based adaptation strategies mentioned above, 65.5% of respondents also depend on loan from relatives, 70.6% on renting/selling own farm land, and 22.3% involve in different handcrafts and local beer making to cope, adapt, and live with the impacts of climate change in order to survive their family as another option. The FGD result also indicated that the households engaged on labor selling and renting their farm lands as a means of alternative to pass the food shortage gaps or to live with unexpected happening of climate-induced shocks/stress in the area.

However, it was indicated by respondents that selling or renting of own farm land for the long period of time may affect the monetary values of that land after a long period of time. Likewise, the existence of this problem was also reported by the FGD and KII participants, they stated that when the Gumuz farmers rent their own lands to other ethnic groups who have an experience in oxen farming they use the land intensively to maximize its return. However, they indicated that it is very difficult for Gumuz farmers to manage the land after renting since its pest and weed infestation increase which requires additional cost and labor to maximize the production to which the Gumuz do not have the experience and capacity to afford. Thought it do not have a legal ground, the Gumuz farmers prefer selling of their lands than renting; this may affect their adaptation ability after a long period of time.

In addition, 74.4% of respondents depend on their better-off relatives to sustain their subsistence life (Table 6). As indicated in the Table 5, looking after relatives for support is negatively correlated with collecting edible wild plants and positively correlated among selling of available household materials or equipments and collecting and selling different wood and wood products with 0.01% significant level. From this it is understood that farmers seek for support from their relatives after exhausting edible wild plants and actively engaged on selling of his/her household materials, and labor-intensive wood and wood products. The practice may have resulted in loss of self-esteems and ability besides affecting the capacity of their relatives too. The correlation result in Table 5 also revealed that the community considers hunting of wild animals as an alternative means even if it does not replace the food demands of the community in the area. In consistence with the declining trends of vegetation cover as a result of expansion in large agricultural investment, currently it is not easy for the community to hunt wild animals as they accustomed before, though it is positively correlated with other autonomous adaptation options except getting support from better-off relatives and collecting and selling of different wood and wood products.

Selling of available household materials/equipment positively correlated with collecting and selling different wood and wood products ($P<0.000$). Also getting support from better-off families affirmatively correlated with collecting edible wild plants and other products, while most of the others correlated negatively. This correlation synergy results indicated that most of "the traditional/autonomous adaptation strategies are not a choice of the community rather it is a means of survival." For instance, the correlation selling of animals are not significantly correlated with collecting edible wild plants and selling/renting of farm lands as indicated in Table 5. This means, when they sell animals and get money to adapt/cope with observed shocks/stress at a given time the household may be not encouraged in collecting of edible plants, hunting, and others. Hence, the correlation result indicated that the community experienced for ages traditional/autonomous adaptation strategies mentioned that are not a choice of the community, rather a means of survival without which their survival is in question. Hence, this analysis result accentuates the importance and timing of introducing the planned adaptation/copping strategies that influenced the community both by the government and/or other development actors.

However, since the community depends on their traditional means for ages, it is very difficult to influence them overnight to accustom the planned/policy-driven strategies; for instance, to transform from hoe farming to oxen farming or other alternative mechanism. Hence, the current situation of the community and the survey findings vigilant the policymakers to consider the subsistence means of the community while making a decision at macro-level to utilize the forest land of the area by providing for large-scale investors, where the life of the community depends on, to bring a better development options in the country as a whole.

Interaction of Policy-Driven and Autonomous Community Resilience to Climate Variability Adaptation and Mitigation

The Gumuz community has been excluded from much of the social and economic activities of the main stream society (Abbute 2002). However, there seem to be some changes in this regard in recent years. In view of this, 47.5% of the respondent households are not aware of having the right to get access to social services. This is despite the fact that our further probing in FGDs and individual interviews showed that their participation does not actually demonstrate equality in the complete sense of the term in any forms of development activities conducted in the area. The development effort seems a top-down approach which overlooks the consent of the beneficiary community. This may have questioned the ownership issues of the development effort under ways in the area. Supporting this statement, Sperling and Szekely (2005) revealed that striking the right balance between top-down command-and-control approaches offer stability over the short term but reduced long-term resilience.

Although limited, improved agricultural practices and technologies, and resettlement programs are the major planed/policy-driven adaptation options promoted to build the resilience of farmers in the areas. Hence, its interaction with community-based adaptation option has discussed.

Agricultural Practice and Technologies: It was generally agreed that the policy responses to climate change should support and enhance indigenous resilience (UNISDR 2009). In this regards, only 42 households in Guba (both in Mankush and Aysid kebeles), 37 households in Dangur, and 20 households in Debate woredas of surveyed kebeleswere reported as they used improved agricultural technologies since the surveyed period. In this regards, Table 7 shows comparison of the actual observed number of farmers started using of agricultural inputs and animals for farming activities with the total number of households in the sample kebeles. This indicated that only a very small number of farmers, 7.7% in Debate, 16.2% in Dangur, and 15.4% in Guba, started using animals for agricultural farm activities. Hence, the result suggest that revisiting of promotional approaches/system in the way that the Gumuz community awareness level and capacity developed to adopt and use introduced technologies as a planned adaptation options to cope, adapt, and live with the observed and perceived climate variability–induced shocks/stress is indispensable.

In addition, capacity building trainings and awareness rising programs were reported as key intervention underway by the government to improve farmers' perceptions towards improved agricultural technologies. However, 66.4% of the respondents reported as they lack awareness on the importance of agricultural technologies, while 11.2% complained unfairness of the available inputs and 15.4% appreciated the existing natural soil fertility than using the artificial once. FGD participants also indicated that the community also lacks their needs of improved and drought-resistant food crop varieties that require short period of time. In this case the existence of mismatching between the needs of the community to adapt the impact of climate change and the promoted agricultural technologies is acknowledged. For instance, the community needs locally adaptive drought resis-tance and high-productive food crop varieties than inorganic fertilizer. Indeed, the FGD and KII participants believe that the voices of the community are often unheard. Also they complained about the price of agricultural inputs, which they attributed to difficulties in reaching the poor community groups which is highly vulnerable and/or susceptible to the impacts of climate-induced shocks due to their low adaptive capacity.

Kebele/local-level development agents believed that although a limited effort has been exerted by nongovernmental organization, namely Canadian Physician for Aid

Table 7 Proportion of households who participated in new agricultural technologies

S.no	Name of sampled woreda	Sampled kebele	Total household	No of household participated in new agricultural technologies (Agricultural inputs)	%
1	Debate	Gessess	260	20	7.7
2	Dangur	Dibate	228	37	16.2
3	Guba	Mankush zuriya	107	14	13.1
		Aysid	224	37	16.5

and Relief (CPAR) and World Vision (WV) Ethiopia to change the deep-rooted saving culture problem of the Gumuz community in the area through organizing them into saving and credit groups, the involvement of local-level government structure is very limited to scale up this effort besides development of their awareness at all level to diversify means of their livelihood. On top of this, among introduced major agricultural technologies only 78% use veterinary services followed by pesticides (15%) and 6.6% ploughing by animals (Table 8). Though maximum efforts have been made by development actors at all level, only 6.6% used ploughing animals for farming practices.

Most of policy-driven agricultural technologies require intensive farming which do not fit with traditional Gumuz farming practice. They alleged that if they produce agricultural crops more than 2 years on a given plots of land continuously the weed and pest infestation increase which requires additional labor and cost. This also needs ploughing by animals than hoe farming where the Gumuz farmers do not have experience, capacity as well as willingness to adopt. In addition, using of a given plots of land for more than 2 years (for 3 and 4 years) requires seed every year which may also costs the farmers every year. Traditionally, once they sow on a given plots of lands they harvest for consecutive 2 years especially of food crops (sorghum). Although it needs further research, the practice seems to have a considerable role in reducing GHG emission through carbon sequestration.

Metekel zone administrator confirmed the existence of strong tension in breaking the traditional means and adopting the new practice. Shifting of hoe farming to animal farming is not achieved overnight. Enforcement of the existing regional land use polices to implement accordingly in the way that it reduces the existing population pressure on forest resources needs to be capitalized. Also the existence of illegal land grabbing in the zone which were initially entered along with legally registered large-scale agricultural investors as a daily laborer worker needs to be checked and corrected. Currently at local government level they considered the investors as a major cause for devaluation of the traditional Gumuz community means of land management and for the observed poor working culture of the indigenous/Gumuz community. The illegal land holders encourage/initiate the Gumuz people to sell, rent, and provide their lands for share cropping than own farm. In addition, the administrator acknowledged as one of the bottlenecks of providing lands for large agricultural investment in the area. Currently the Gumuz communities especially the youngsters considered themselves as investors, predominantly in Debate and Guba woredas as witnessed by FGD participants.

Table 8 Major agricultural technologies introduced and used by the farmers

Technologies	Yes		No	
	count	%	Count	%
Veterinary service	135	78	38	22
Pesticides	26	15	147	85
Artificial insemination	2	1.2	171	98.8
Ploughing by oxen/tractor (Agricultural implements costs)	41	6.6	57	93.4

Because socially in their community they have a right to use their clan land; but currently they started selling, renting, and give for share cropping their clan lands which is large in size/coverage even though traditionally land is not sold under any circumstances either inside or outside the clan of the Gumuz community (Patrick Wallmark 1981).

Hence, the traditional clan land management systems of the Gumuz community is changing with time which may be as a result of population pressures and the extreme climate conditions that forced the highlander immigration to find an alternative means of fulfilling their basic needs and/or provision of forest lands for large-scale agricultural investment.

To this end, it is noted that the community started questioning the government to hear their voice and close a reasonable size of forest land areas for the community. "Head of agriculture and rural development office of Guba woreda confirmed that the land allocation and provision systems for large investors in the area were not considering the local context. Also pointed as, the woreda/distrct officials are exerting their maximum effort in collaboration with zonal and regional government to create a situation in which the local voice should be heard and considered during allocation of large forest lands for investment works. As an adaptation option, promotion of technology transformation from the investor to farmers, establishment of local-level research centers, introduction of appropriate veterinary services, income source diversification through appropriate management of livestock's, and promotion and development of appropriate plan in which the available water sources like Beless, Ayma, and Abay rivers used for irrigation as a major development directions were set at woreda level."

Generally, all policy-driven technologies seem to improve the knowledge and capacity of the community to cope, absorb, and adopt with stress and shocks beside its contribution to increase emission from all land use practices unlike the traditional means. Its rate of promotion is overwhelmed by the current rate of climate variability–induced impacts that threatened the life of the Gumuz people in the areas.

Resettlement: With the objective to introduce improved technology and infrastructure for ensuring food security and to change the lifestyle of the Gumuz ethnic groups, villegization program was initiated by the government and implemented in the area since the year 2012. Most of the farmers were resettled into centralized villages where the water is available or planned to be accessed. However, the FGD participants testified that the community prefers to live in their traditional villages as practiced by their ancestors. Hundred percent of the resettled households whispered as they got no remittance payments made to the community for the resettlement programs. They also mentioned as they face difficulties in identifying what and where to get their needs from forest resources in the new areas since there is no other alternative options to support their ways of life. They also mentioned reduction of forest resources as a result of pressures induced by resettlement program that threatened their life. They reported that women sometimes leave newborns at home to go the whole days for collecting forest products; otherwise there will be no food for the family as a whole. Women were also got difficulties in getting resources to produce different handcrafts, the important economic activity in the area, especially for the women, both for home use and market. Besides its importance in accessing basic socioeconomic services, the program seems as it threatened the community traditional means of copping and living with the current and perceived climate variability–induced shocks/stress. In this regard, they are urging revisiting of the program and development of strategic plans that accommodate the interest of the community in supporting their ways of life means in their new areas.

The three key issues that would affect the successful implementation of the planned/policy-driven adaptations options which need the attentions of policymakers when making decision to use the area for other development purpose are identified. The first was the level of technical assistance, given the limited experience and lack of local language knowledge of the government extension workers in the study zone. As a consequence, extension agents do not have skills to train and influence farmers on the adaptations that could be implemented now, or to provide adequate technical support for routine agricultural production practices. Technical assistance from NGOs and community-based extension emphasize that farmer-to-farmer knowledge exchange supported by targeted technical assistance is needed to address the lack of formal extension services.

The second was poor working and saving culture of the community. Since they are highly dependent on forest and forest products they do not have even storage facility for their produces. They store in the farm at field and outside of their village. Also the mens do not have commitment to be engaged on agricultural farm activities alone with his family but women. Their farm activities have been conducted through a self-help group, locally named as "debbo." Debbo requires preparation of local drinks the self-help members to conduct agricultural activities for debbo owner. All arrangement and preparation of necessary materials for debbo is the duty of women in the community. Especially during clearing of forest lands, weeding, harvesting, and threshing of crops debbo is must. The farm land size and amount of crop harvested will be also determined by the number of the debbo that a given household would be managed.

Finally, Gumuz farmers lack experience in oxen/animal farming and conducting intensive farm operations besides lack of capacity to have animal or any other small technologies for agricultural farm operation purpose. Malone (2009) indicated that the adaptation process requires the capacity to learn from previous experiences to cope with current climate, and to apply these lessons to cope with future climate, including surprises. Hence, introduction of planned/policy-driven adaptation strategies based on the existing community-based adaptation options that influence the community is indispensable to cope and adapt with extreme weather- and drought-induced shocks/stress in a sustainable way through building their resilience. In this regard, any policy-driven adaptation options that address the above-mentioned key elements are required to improve the adaptive capacity of the Gumuz community in the area.

Conclusions and Recommendations of the Chapter

Conclusion

The focus areas of this chapter, Metekel zone, northwest Ethiopia, scrutinized how the Gumuz communities adapt the adverse effects of climate variability and investigated determinants of the use of identified adaptation strategies/options in the area.

The farmers perceived the declining trends of crop yield, livestock production, and the benefits that community gets from forest resources for the last 5 years. Although it is different among respondents' wealth categories, the existence of food shortage gaps in the community that extends up to 7 months has been identified. The situation is worse in the case of poor community members.

In response to climate change impacts, the Gumuz communities have been practicing different adaptation strategies. Many of the autonomous adaptation responses to the impacts of climate variabilitys/changes include collecting of wild edible plants and other forest products, selling livestock (particularly when they are not likely to get a good price), hunting, renting/selling of farm animals, selling of available household materials and/or equipments, getting support/relay on better-off relatives, and collecting and selling of different wood and wood products (Table 6). In addition, expansion of farm land, usage of drought-resistance crop varieties, replanting/sowing, and changing of sowing calendar are among reported strategies used in response to the impact of climate variability/change. However, the autonomous/community-based adaptation mechanisms have been declining both in type and amount. This is mainly due to provision of forest lands for large-scale agricultural investments and the progressively increasing population pressure. The few households are able to improve their farming practices, for example, in using improved agricultural inputs. Lack of awareness on improved agricultural practices, poor saving, and working culture and dependency on natural resources to fulfill their food shortage gaps were known as the main constraints to adopting these practices. These further resulted in declining of community resilience to the level of crisis which requires direct intervention and support. This calls for a planned action to ensure households meet their consumption needs.

Recommendations

- Unless communities actively engage in reflexive learning processes about the causes of systemic changes and the links between local and global processes, there is a risk that community resilience becomes nothing more than an illusion. Hence, introduction of awareness raising programs that improve working and saving culture and climate information appear to be important mechanisms as they support the adoption of several adaptation strategies that build community resilience. Special skills training program is keenly important for the local community so that they will participate and benefit from both government- and private sectors–initiated development programs which are underway in the area.
- The capacity of the community to adopt climate change by using autonomous adaptation is degrading. Therefore, greater effort is needed to increase the resilience of households to cope and adapt with climate variability, through maintaining their autonomous means systematically, and social safety net programming till the community transformed from their autonomous/community-based means to improved policy-driven adaptation mechanisms and adopt the

practice besides encouraging accumulation of assets and wealth through diversifying locally adaptive options.

- Given the effect of climate change on crop yields, animal production, and food availability, planned action by government is needed to ensure households meet their consumption needs. This may take the form of protecting the existing natural forest ecosystem where the community can get their day-to-day needs; for example, promotion of bamboo plantation using different plantation strategies since the community has a special attachment with the species to maintain their food security. Other public actions that would increase access to weather insurance for cash crops, of those currently adopted by the community (sesame production) and creating market linkage, early-warning preparedness, and increasing food stockpiles to be used during poor production years need to be considered.

- Besides the available potentials and the existing policy direction at national level, irrigation development interventions are overlooked both at individual and community level in the area. This suggests that investments in irrigation infrastructure would help farmers to engage in higher value crops, thereby increasing farm revenues besides creating alternative farm production for their subsistence life. There is also an identified need for greater investments in promoting appropriate agricultural extension services, locally adaptive and affordable technologies, and accessing local-level research institution to support the effort of the community in availing locally adaptive food crops variety to improve the future well-being of the community in the area.

References

Abbute W-S (2002) Gumuz and Highland Resettlers: differing strategies of livelihood and ethnic relations in Metekel, Northwestern Ethiopia. PhD Dissertation, University of Gottingen, Gottingen

Alinovi L, Mane E, Romano D (2009) Measuring household resilience to food insecurity: application to Palestinian households. Working Paper. EC-FAO Food Security Programme

Anonymous (2004) Benishangul-Gumuz Regional State Rural Development Coordination Office (2004) Three-year Strategic Plan (2003/04–2005/06)

Anonymous (2011) Benishangul-Gumuz Regional State Food Security and Economic Growth, Canadian based Non-Governmental Organization Five Years Project implementation Plan (2011–2016)

Berhe MG, Butera J-B (2012) Climate change and pastoralism: traditional coping mechanisms and conflict in the horn of Africa. Institute for Peace and Security Studies. Africa Programme, Addis Ababa

Emerta AA (2013) Climate change, growth and poverty in Ethiopia. Working Paper no 3. CCAPS

Esayas A (2003) Soils of Pawe Agricultural Research Center. National Soil Research Center Ethiopian Agricultural Research Organization Technical paper no 78. 23pp

Food Security Strategy, Benishangul-Gumuz, Ethiopia (October 2004) Asossa (unpublished) Global Environmental Change 16:253–267. https://doi.org/10.1016/j.gloenvcha.2006.04.002

IPCC (2007) Climate change 2007: impacts, adaptation and vulnerability. Contribution of Working Group II to the Fourth Assessment Report of the Intergovernmental Panel on Climate Change,

Annex I (eds: Parry ML, Canziani OF, Palutikof JP, van der Linden PJ, Hanson CE). Cambridge University Press, Cambridge, UK, 976pp

Nyong A, Adesina F, Elash OB (2007) The value of indigenous knowledge in climate change mitigation and adaptation strategies in the African Sahel. Mitig Adap Strat Glob Change 12:787–797

Seo SN, Mendelsohn R (2008) Measuring impacts and adaptations to climate change: a structural Ricardian model of African livestock management. Agricultural Economics, p 38

UNISDR (2009) Terminology: basic terms of disaster risk reduction and IISD et al. 2007. Community-based Risk Screening – Adaptation and Livelihoods (CRiSTAL) User's Manual, Version 3.0

Wallmark P (1981) The Bega (Gumuz) of Wollega: Agriculture and Subsistence; in people and cultures of Ethio-Sudan Borderlands (ed: Bender ML). African Studies Center, Michigan State University, East Lansing, p 79

World Food Program (WFP), Food and Agriculture Organization (FAO) (2012) WFP/FAO Shock Impact Simulation Model for food security analysis and monitoring

7

Attaining Food Security in the Wake of Climatic Risks: Lessons from the Delta State of Nigeria

Eromose E. Ebhuoma

Contents

Abstract

Climate variability and change have undermined the poor rural households' ability in sub-Saharan Africa (SSA) to engage in food production effectively – which comprises their primary source of livelihood – partly because it is predominantly rain-fed. Notwithstanding, the rural poor are not docile victims to climatic risks. They actively seek innovative ways to utilize their bundle of assets to reduce the negative effects of climatic risks to ensure household food security. Bundle of assets comprise the financial, human, physical, social, and natural assets owned by, or easily accessible to, an individual. Drawing on primary data obtained qualitatively in the Delta State of Nigeria, this chapter analyzes

E. E. Ebhuoma (✉)
College of Agriculture and Environmental Sciences, Department of Environmental Sciences, University of South Africa (UNISA), Johannesburg, South Africa
e-mail: ebhuoee@unisa.ac.za

how Indigenous farmers utilize their bundle of assets to grow their food in the face of a rapidly changing climate. The results indicate that human and social assets played crucial roles in facilitating household food security. Also, social assets facilitated the procurement of other assets necessary to ensure continuity in food production, albeit farmers continue to live under the global poverty line. This chapter critically discusses the implications of these findings in relation to the attainment of both the first and second Sustainable Development Goals (no poverty and zero hunger) by 2030 in the Delta State.

Keywords

Assets · Climate change · Adaptation · Food security · Indigenous farmers; Nigeria

Introduction

Climate variability and change have adversely affected various sectors of the global economy including health (Ebhuoma and Gebreslasie 2016), transportation (Jaroszweski et al. 2010), and tourism (Fitchett et al. 2017). However, no sector has been severely affected like agriculture, especially in the developing world (Intergovernmental Panel on Climate Change (IPCC) 2014). This is primary because the agricultural practices embarked upon by poor rural households are extensively dependent on rainfall (Conway and Schipper 2011). Consequently, the slightest deviation of weather patterns from the normal can subject most of the rural poor in developing countries to excruciating poverty and misery due to their inability to obtain their livelihood from food production (IPCC 2014). Furthermore, the vulnerability of the rural poor to climatic risks is exacerbated by weak institutions and agricultural policies, deficiency of social safety nets, inability to purchase farm insurance, and low levels of education (Perez et al. 2015).

In Nigeria, for example, agriculture contributes about 20% to its gross domestic product (GDP), making it next in line to the country's mainstay after crude oil (National Bureau of Statistics (NBS) 2014). In the last two decades, however, climate variability and change have wreaked havoc in various farming communities, especially in the Delta State where 90% of rural households are actively engaged in food production (Ifeanyi-obi et al. 2012). Climatic risks have become a huge cause for concern among the rural poor due to growing uncertainty regarding anticipated food productivity and outputs (Mavhura et al. 2013; Nelson et al. 2014). Despite the increased climatic risks that the rural poor in the Delta State and other parts of sub-Saharan Africa (SSA) are besieged by, they are not docile victims to these threats.

The poor, as Moser (2011) argue, are actively and consistently seeking innovative ways to utilize, modify, and adapt their bundle of assets or capital to reduce the negative effects of climatic risks on their livelihood. Bundle of assets comprises the financial, human, physical, social, and natural assets (Table 1) owned by or easily accessible to an individual. The focus on assets is crucial to facilitating the

Table 1 Definition of bundle of assets

Asset or capital	Definition
Physical	This includes equipment, infrastructures such as road networks, and other productive resources owned by individuals, households, communities, or the country itself
Financial	This refers to financial resources available and easily accessible to individuals, which includes loan, access to credits and savings in a bank or any other financial institutions
Human	This refers to the level of education, skills, health status, and nutrition of individuals. Labor is closely associated with human capital investments. Health statuses of individuals impact either positively or negatively on their ability to work, while skill and level of education is crucial because it influences individuals return from labor
Social	This refers to the norms, rules, obligations, mutuality, and trust embedded in social relations, social structures, and societies' institutional disposition
Natural	This refers to the atmosphere, land, minerals, forests, water, and wetlands. For the rural poor, land is an essential asset.

Sources: Bebbington (1999); Moser and Satterthwaite (2008); Moser (2011)

identification of entry points to inject tailored policy interventions that are necessary to build and fortify the adaptive capacity and resilience of the rural poor (Moser 2011; Moser and Stein 2011). As documented by Moser (2011), individuals are not docile victims but possess resources that they can draw upon in times of crisis. Thus, identifying and strengthening these resources is crucial for the poor to be able to hold their own in times of crisis such as climate variability and change by deploying their available resources to ensure food security.

In the wake of a rapidly changing climate, the injection of tailored policy interventions is desperately needed to scale up food production in SSA and facilitate the actualization of Sustainable Development Goals (SDGs) 1 and 2 (no poverty and zero hunger) by 2030. Against this background, this chapter analyzes the ways in which Indigenous farmers in Igbide, Uzere, and Olomoro communities in the Delta State of Nigeria utilize their bundle of assets to grow their food in the face of a rapidly changing climate. Indigenous, in this context, refers to people that possess a peculiar culture and knowledge distinct to their community that have been examined with real-life scenarios (Ebhuoma 2020).

Research Methodology

The chapter is based on primary data obtained in Olomoro, Igbide, and Uzere communities situated in *Isoko* south local government area (ISLGA) of the Delta State in Nigeria (Fig. 1). The mean annual rainfall in the Delta State is between 2500 to 3000 mm (Adejuwon 2011). Both Igbide and Uzere are low-lying, while Olomoro comprises both high- and low-lying areas. Due to annual heavy rainfall events, the low-lying farmlands are submerged from June to the last week in October.

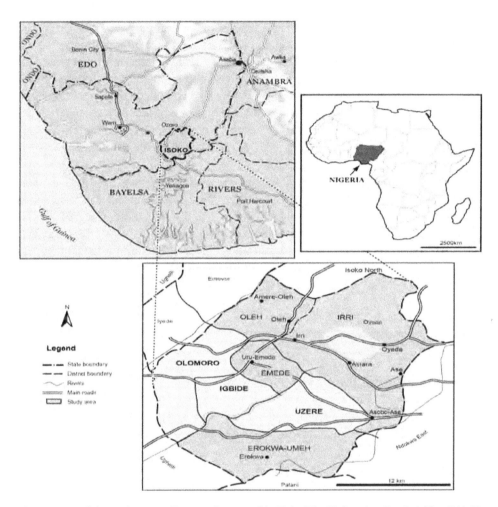

Fig. 1 Map of the study areas. (Source: Cartographic Unit, Wits University, South Africa (2016))

Omohode's (2012) documentation, following the 2012 flood disaster that severely affected most States in Nigeria, influenced the choice of these communities. He highlighted that most low-lying communities in ISLGA were completely submerged, making the area resemble emergency oceans when viewed from a distance. Thus, unpacking the ways in which Indigenous farmers in these communities engage with their bundle of assets will provide valuable insights regarding how vulnerable people grow their food in the face of climatic risks.

The communities are homogeneous in nature. For instance, *Isoko*, an Indigenous language, is the local dialect spoken. Also, small-scale farming is the major economic driver of these three communities, with the women at the helm of the practice. While some men assist their wives to produce food, they are mostly involved in fishing. In terms of food production, cassava and groundnut are the predominant staples cultivated annually. Cassava makes up approximately 65% of the total caloric intake in each community. Other cultivated crops include cocoyam, potato, pepper, and plantain. With the exception of cassava which requires a minimum of 6 months to reach maturity, the other crops can be harvested 3 months after planting.

Focus group discussions (FGDs) and semi-structured interviews were used to obtain primary data. Thirty-five FGDs and four one-to-one, semi-structured interviews (two in Olomoro, one in Igbide and Uzere) were conducted between June and October 2015. Of the 35 focus groups, 24 were made up of female respondents; five comprised male respondents, while six were made of both male and female respondents. Respondents in each FGD varied between 3 and 12 respondents aged between 20 and 85.

Respondents were identified using purposive sampling based on age, gender, those who have been farming in the study areas for a minimum of 10 years, those whose household assets and livelihoods were severely affected by the 2012 flood disaster, and those that grow their food on low-lying farmlands. Key informants who have lived in each community for over 40 years and an agricultural extension worker facilitated the recruitment of eligible respondents. Primary data retrieved were analyzed using the thematic analysis technique.

Findings

Livelihood Vulnerability to Climatic Risks

Respondents pinpointed heavy rainfall events – which resulted in seasonal flooding of low-lying farmlands annually – as the worst weather conditions that undermined effective food production through farmers' inability to maximize their natural capital. In this regard, a respondent from Uzere, in his 80s, stated:

> We are constrained to practice seasonal planting due to flooding which must occur on our farmland annually. Consequently, we must harvest all our cultivated cassava before our farmland gets inundated. This usually worsens food insecurity in times of poor harvest.... This is the advantage farmers in neighboring communities who cultivate on high ground have over us. They do not lack *garri* (processed cassava) throughout the year.

Seasonal flooding restricts farming for 8 months annually, which has implications for the amount of food farmers are able to produce annually. The second weather conditions that adversely impacted food production are rising temperatures, especially between February and April. On the one hand, respondents aged 40 years and below revealed that the weather has become warmer in the last decade. On the other hand, the elderly respondents (50 years and above) argued that the rise in temperature began in the early 1980s. Both groups unanimously acknowledged that in the last 5 years, temperatures between February and April have become abnormally high in the afternoons. This has undermined their ability to work effectively on their farmlands. From the respondents' viewpoint, the adverse effects of the rising temperature are evident in groundnut production as they now harvest empty pods more frequently than in previous times. A respondent from Igbide, in her 50s, asserted:

> The sunlight during the months of February and March is really terrible and planting during those months is very difficult. Groundnut is the crop that is seriously affected because it is does not require intense sunlight for optimal productivity.

Approximately two-thirds of the respondents stated that the change in weather conditions is due to God's making and supernatural forces. When probed about the role of humans in contributing to climate change, most debunked the claim. To concretize this viewpoint, female respondents in a FGD in Igbide unanimously agreed that *the change in weather, humans have nothing to do with it; it is solely the making of God*. However, the youths attributed the vagaries of weather to the increased rate of deforestation carried out by farmers to obtain firewood. Also, few elderly respondents revealed that the rising temperatures are due to continuous gas flaring activities by Shell's crude oil exploration activities for over 40 years. These three communities have about 62 oil wells that Shell drilled oil from before selling all its oil wells in these communities to the Integrated Data Services Limited (IDSL), a subsidiary of Nigerian national petroleum corporation (NNPC), in 2014. In this light, a male respondent from Uzere, in his 50s, commented:

> This community is particularly known for farming. But since the early 1980s, the quality of both cassava and groundnut produced has reduced significantly. This is due to Shell's oil exploration activities. Most of the youths now engage in off-farm activities because farming can no longer foot their bills.

Most elderly respondents attributed the poor starch content of the *garri* (processed cassava) they produce to crude oil exploration activities. They lamented that the oil exploration had compromised their soil's nutrients, which in turn has affected the nutritional value of the *garri* produced, especially when compared to the produce harvested in the 1980s. However, only a few respondents highlighted farmers' inability to engage in bush fallowing, due to increased demand for land stemming from sporadic population growth, as an added factor that has facilitated the reduction in quality of food produced.

Assets and Food Production Nexus

Due to the annual seasonal floods, farmers employ their human capital to produce cassava on their low-lying farmland through an Indigenous strategy referred to as *elelame* (follow the water). The other cultivated crops – cocoyam, potato, pepper, and plantain – are produced using the early rains, which usually begins between February and March and last till the end of May. The water strategy commences as soon as the floodwater starts to recede the farmland, usually in November. The farmers' plant their cassava stems on the part of the soil that is visible and moist. They replicate this process until the floodwaters have completely dried up from their farmland. The planting process usually ends between the second and third week in December.

The following year, between June and August, when the rain is heavy and continues to fall consistently, they start harvesting their produce. The decision regarding where to commence harvesting is hinged on their human capital informed by their Indigenous knowledge, as they know the precise portion of their farmland that will be submerged at the earliest. Thus, they do not harvest all their farm produce simultaneously. The crop closest to where the inundation will commence are harvested first. The reason for not harvesting all the produce at the same time is because the longer cassava remains in the soil, the bulkier they get. Also, labor shortage is another factor that contributes to adopting this harvesting strategy. Thereafter, usually within a week, they would return – pending on the consistency of rainfall – to their farmland to employ a similar process to harvest the other produce. After harvesting all their produce, they preserve the cassava stems on their inundated farmland by constructing temporary structures to use them for food production in the next planting season (Fig. 2).

To ensure they have *garri* to eat all year round, they utilize their human capital to process the harvested cassava as well as store it properly. Respondents explained that after the necessary procedures have been implemented, which entail peeling, soaking the cassava in water for several hours, drying the soaked tubers and blending into powdered form, it is fried with little palm oil to an overly dried state. After cooling down, the *garri* is preserved in airtight sack bags. Thereafter, a wooden structure is constructed and the sack bags placed on top of it. The fundamental reason for suspending the sack bags from the ground is to prevent the *garri* from going bad through mold formation.

Fig. 2 Indigenous technique used to preserve cassava stem on low-lying farmland. (Photograph: John Ayiko (2015))

It is noteworthy to mention that some farmers rent farmland, a natural capital, to grow their food. Because most farmers lack financial capital during the planting season, social capital plays a vital role in this regard only for trustworthy individuals. As a respondent from Uzere, in her 40s, highlighted:

> Most farmers lack finances during the planting season. Consequently, only trustworthy individuals are privileged to get farm plots leased to them without having to pay the agreed sum upfront. Often times, they pay the landowners after harvesting and traded some of the produce.

Also, some farmers – due to a shortage of household labor and lack of capacity to hire laborers – drew on their social capital to acquire human capital to facilitate the harvesting of farm produce before the occurrence of the seasonal flooding. Specifically, some farmers depend on neighbors, relatives, and friends to accelerate the harvesting process to avert the possibility of some of the produced cassava from decaying. Furthermore, social capital catalyzed the procurement of financial capital. This is particularly useful as most farmers have been unable to benefit from several loan schemes afforded by the Delta State government against the backdrop of the farm loans being disbursed consistently for over 10 years (United Nations Development Program (UNDP) 2014). Some highlighted that they only hear of farm loans after the application process had closed, a state of affairs which was largely attributed to nepotism.

Although microfinance banks (MFB) in the Delta State have been given directives to provide the rural poor with farm loans, the inability to provide collateral matching the value of the loan sought after or a guarantor with valuable assets has hampered farmers' ability to secure such loans. As respondents in a female-only FGD in Olomoro bemoaned:

> Loans exist that could reduce some of the challenges we undergo as farmers, but due to the fact that there is nobody to stand as a guarantor [lack of social capital], they have not been able to harness such opportunities.

Since farmers' annual earnings from food production (between 137 USD to 219 USD) are inadequate to secure their livelihood objectives, they utilize their social capital to temper the financial drought. This is achieved by some community members coming together to form a small group where the prior agreed monetary contributions are made weekly to a trustworthy individual. At the end of each month, the total sum is given to a group member, hinged on prearrangement. This scheme, referred to as *Osusu*, is useful in ensuring that farmers can purchase items necessary for food production.

Households Still Living Below the Global Poverty Line

Despite farmers' skillful utilization of their meager bundle of assets at their disposal to ensure continuity in food production, the majority still live under the global poverty line of less than $US 1.90 a day (Livingston et al. 2011). A fundamental reason for this is due to the low financial gains made from the sale of *garri* underpinned by its inferior quality when compared with those produced in neighboring communities' void of oil exploration activities. Thus, they are "forced" to market their produce at a much-reduced price.

Another factor that impeded farmers from transcending living above the global poverty line is due to the exorbitant interest rate money is borrowed from unregulated bodies such as informal meeting groups and money lenders. Respondents highlighted that not knowing influential people, underpinned by lack of social capital, to act as guarantors to co-sign the credit agreement to access farm loans from MFB is a pull factor toward securing loans from unregulated sources. As some respondents explain, this is prevalent during the planting season as farmers often run out of money having addressed other pressing issues such as paying for both children's tuition fees and levies attached to social responsibility. Thus, farmers are left with no feasible alternative but to obtain loans from "financial predators" as their requirements are less demanding. While the loan obtained enables farmers to produce their food, it proved counterproductive in terms of evading the poverty maze. For example, if a farmer borrows 50 USD for 6 months, the farmer is required to refund the loan with a whopping 40% interest. This is testament to the fact that the drive to become food secure pushes farmers to do anything within their powers to achieve the objective, regardless of the long-term consequences.

The financial predators are well knowledgeable on the importance of farm loans in ensuring household food security. As a result, they are unwilling to water down their terms and conditions. In this regard, a farmer from Igbide, in his 50s, explained:

> Without loans, some farmers cannot grow food. After these farmers secure loan from non-government bodies, grow their food and sold some of the produce to refund the loan, most of the time, they are left with little or nothing for the next planting season. The only choice they have is to go back to secure loans from the group that lend them money previously. This is the survival tactic of some farmers in this community.

In fact, the inability to access loan is a catalyst that has made some farmers to engage in off-farm activities. Another factor that compromised effective food production was the lack of physical capital, especially for farmers with access to large hectares of land enough to engage in commercial farming. For example, farmers' inability to access farm machinery dampened their fight to transcend the boundaries of a subsistence farmer. A male respondent in Uzere stated that while the Delta State government usually provides farm equipment for farmers, "it never gets to them." Instead, the equipment is "always hijacked" by influential politicians and close associates of key politicians in the Delta State. In addition, the unavailability of rice milling machines has prevented farmers from producing rice. Few elderly respondents (50 years and above) in Igbide revealed:

> In the 1960s, they were actively involved in rice production because of the swampy nature of their farmlands, and rice milling machines provided by the government. But since the 1970s till date, no provision has been made to provide rice milling machines for farmers. As a result, rice cultivation has been abandoned.

Finally, the farmers lamented bitterly that despite the enormous contributions their communities have made to the nation's foreign revenue for over four decades, their communities have remained shockingly underdeveloped. The lack of good road networks within each community, for example, erodes the financial capital of some farmers, albeit insidiously. To illustrate, during the rainy seasons, it can be challenging for motorists to navigate their way through their community due to countless potholes. This makes accessibility to markets where they have to sell some of their farm produce an exasperating venture.

Discussion

The adverse effects of climatic risks are palpable in Igbide, Uzere, and Olomoro communities in the Delta State of Nigeria. They have manifested in the form of heavy rainfall events (Ifeanyi-obi et al. 2012), which leads to seasonal flooding of low-lying farmlands, and rising temperatures (Ike and Ezeafulukwe 2015). While these climatic variables have undermined food production, oil exploration activities have aggravated farmers' woes. By significantly degrading soil's nutrients, oil exploration activities have adversely compromised the quality of food produced. This assertion is corroborated by research findings that have also emerged from the Delta State (Ererobe 2009; Elum et al. 2016). Nonetheless, the findings can be disaggregated into two key points.

First, farmers' perception of climate change is underpinned by religious framing. Similar findings have been recorded in Botswana (Spear et al. 2019), Mali (Bell 2014), Nigeria (Jellason et al. 2020), South Africa (Okem and Bracking 2019), and Zimbabwe (Moyo et al. 2012), respectively. Attributing the cause of climate change to oil exploration activities, God and other supernatural forces as well as disentangling their lifestyle activity – deforestation – as a contributing factor seems the logical way for people to continue with the state of affairs without any ill feelings. Accepting how their lifestyle choices may be contributing to climate change, no matter how insignificant it may seem in comparison to gas flaring, for example, will doubtlessly require behavioral changes. In contrast to studies that show that people highly vulnerable to climate change may be more willing to adopt behavioral changes (Akerlof et al. 2013; Azadi et al. 2019), this may not be feasible for farmers in the Delta State due to their quest to obtain their livelihood by any means necessary.

For instance, to rely on kerosene or gas-fueled stoves for cooking may have substantial financial implications in comparison to firewood. In this light, therefore, the need to sensitize farmers on how their actions are contributing to climate change, including the possible future implications for household food security, is essential. This is primarily because rural households in SSA are expected to be adversely affected by the impacts of future climate change (IPCC 2014). Also, it is necessary to involve religious clergies as key stakeholders in the discourse around climate change mitigation as their beliefs and values have the potential to influence the behaviors of their congregation.

Second, farmers have carved out unique strategies to maximize their meager asset portfolios to produce food despite the increasing threats from climatic risks. The systematic ways in which farmers utilize their human capital to grow food on their low-lying farmlands indicate that farmers are not helpless victims to climatic risks. This is corroborated by findings in Bangladesh (Al Mamun and Al Pavel 2014), Botswana (Motsumi et al. 2012), and Zimbabwe (Mavhura et al. 2013). With the right support and interventions such as providing easy access to loans, machinery, and good road networks, the likelihood that farmers will successfully transcend living above the global poverty line is extremely high. As several studies show (Kochar 1997; Akoijam 2012; Ibrahim and Aliero 2012; Assogba et al. 2017), easy access to government loans remain a major challenge for rural farmers in developing countries. It is documented that the flourishing of exploitative money lenders is due to low priority given to rural credit (Akoijam 2012). Thus, to ensure farmers access farm loans, robust broadcasting of any program through mediums utilized by households to receive vital information are crucial. Otherwise, the persistent dependence on financial predators will continue to flourish, to the detriment of farmers in the Delta State achieving the first SDG.

It should be emphasized that the skillful utilization of social capital to acquire human capital (assistance with cassava harvesting), financial capital (*Osusu*), and natural capital (not paying the rental before cultivating on farmland) indicates that climate adaptation interventions that may cause fragmentation of households should be avoided. For example, suppose the government wants to provide farmlands on higher grounds to farmers to ensure they can produce food all year round. In that case, farmers in the same community should be given land close to one another. This is crucial for the strengthening of social capital, which is essential to facilitating household food security and ensuring that the country is on the trajectory toward achieving the first (no poverty) and second (zero hunger) SDGs by 2030. As Joshi and Aoki (2014) argue, strong social networks influence household's ability to recover from a disaster.

Final Remarks

Climatic risks are making life difficult for the farmers cultivating on the low-lying farmlands in Olomoro, Uzere, and Igbide communities. In responding to these threats, Indigenous farmers skillfully employ their limited bundle of assets to continue producing their food. Specifically, this chapter illustrated how human capital plays a pivotal role in ensuring the production of cassava in the low-lying farmlands, which experiences seasonal flooding annually, through an Indigenous strategy referred to as elelame (follow the water). Also, social capital is a crucial asset in farmers' portfolio through its ability to facilitating the procurement of financial capital through a local scheme called Osusu. Further, it enabled the acquisition of natural capital by allowing trustworthy individuals to renting farmlands and only paying the fee after harvesting and selling some of the produce. Since social capital is overwhelmingly fundamental to the achievement of food security, any scheme meant to assist farmers to adapt more effectively to climatic risks to produce more food must ensure it creates an avenue for strengthening ties among farmers.

This chapter also finds that despite farmers' ability to attain household food security every year, they still live below the global poverty line. A key factor fuelling this state of affairs is primarily due to the inaccessibility of government loans. Consequently, financially strapped farmers are constrained to secure loans from unregulated sources. While it provides a leeway to continue in food production, it is counterproductive due to the high-interest rates attached to the loans. Perhaps, easing the loan acquisition process from MFB may successfully combat this menace. Otherwise, farmers will be unable to weave their way out of poverty. To conclude, until interventions are geared toward ensuring the protection, strengthening, and making the acquisition of assets that play a fundamental role in food production, chances of successfully achieving the first and second SDGs will be slim.

References

Adejuwon JO (2011) A spectral analysis of rainfall in Edo and Delta states (formerly mid-western region), Nigeria. Int J Climatol 31:2365–2370

Akerlof K, Maibach EW, Fitzgerald D, Cedeno AY, Neuman A (2013) Do people 'personally experience' global warming, and if so how, and does it matter? Glob Environ Chang 23(1):81–91

Akoijam SLS (2012) Rural credit: a source of sustainable livelihood of rural India. Int J Soc Econ 40:83–97

Al Mamun MA, Al Pavel MA (2014) Climate change adaptation strategies through indigenous knowledge system: aspect on agro-crop production in the flood prone areas of Bangladesh. Asian J Agric Rural Dev 4(1):42–58

Assogba PN, Kokoye SEH, Yegbemey RN, Djenontin AJ (2017) Determinants of credit access by smallholder farmers in North-East Benin. J Dev Agric Econ 9(8):210–216

Azadi Y, Yazdanpanah M, Mahmoudi H (2019) Understanding smallholder farmers' adaptation behaviours through climate change beliefs, risk perception, trust, and psychological distance: evidence from wheat growers in Iran. J Environ Manag 250:109456. https://doi.org/10.1016/j.jenvman.2019.109456

Bebbington A (1999) Capitals and capabilities: a framework for analyzing peasant viability, rural livelihoods and poverty. World Dev 27:2021–2044

Bell D (2014) Understanding a "broken world": Islam, ritual, and climate change in Mali, West Africa. J Study Relig Nat Cult 8(3):287–306

Conway D, Schipper ELF (2011) Adaptation to climate change in Africa: challenges and opportunities identified from Ethiopia. Glob Environ Chang 21:227–237

Ebhuoma E (2020) A framework for integrating scientific forecasts with indigenous systems of weather forecasting in southern Nigeria. Dev Pract 30:472–484

Ebhuoma O, Gebreslasie M (2016) Remote sensing-driven climatic/environmental variables for modelling malaria transmission in sub-Saharan Africa. Int J Environ Res Public Health 13(6):584. https://doi.org/10.3390/ijerph13060584

Elum ZA, Mopipi K, Henri-Ukoha A (2016) Oil exploitation and its socioeconomic effects on the Niger Delta region of Nigeria. Environ Sci Pollut Res 23:12880–12889

Ererobe M (2009) FG, multinationals and Isoko nation. https://www.vanguardngr.com/2009/08/fg-multinationals-and-isoko-nation/. Accessed 3 Apr 2020

Fitchett JM, Robinson D, Hoogendoorn G (2017) Climate suitability for tourism in South Africa. J Sustain Tour 25(6):851–867

Ibrahim SS, Aliero HM (2012) An analysis of farmers' access to formal credit in the rural areas of Nigeria. Afr J Agric Res 7:6249–6253

Ifeanyi-obi C, Etuk U, Jike-wai O (2012) Climate change, effects and adaptation strategies: implication for agricultural extension system in Nigeria. Greener J Agric Sci 2:53–60

Ike PC, Ezeafulukwe LC (2015) Analysis of coping strategies adopted against climate change by small scale farmers in Delta state, Nigeria. J Nat Sci Res 5:15–24

Intergovernmental Panel on Climate Change (IPCC) (2014) Africa. Intergovernmental Panel on Climate Change. https://www.ipcc.ch/pdf/assessment-report/ar5/wg2/WGIIAR5-Chap22_ FINAL.pdf. Accessed 25 Sept 2020

Jaroszweski D, Chapman L, Petts J (2010) Assessing the potential impact of climate change on transportation: the need for an interdisciplinary approach. J Transp Geogr 18:331–335

Jellason NP, Conway JS, Baines RN (2020) Exploring smallholders' cultural beliefs and their implication for adaptation to climate change in North-Western Nigeria. Soc Sci J. https://doi.org/10.1080/03623319.2020.1774720

Joshi A, Aoki M (2014) The role of social capital and public policy in disaster recovery: a case study of Tamil Nadu state, India. Int J Disaster Risk Reduct 7:100–108

Kochar A (1997) An empirical investigation of rationing constraints in rural credit markets in India. J Dev Econ 53:339–371

Livingston G, Schonberger S, Delaney S (2011) Sub-Saharan Africa: the state of smallholders in agriculture. http://www.ifad.org/documents/10180/78d97354-8d30-466e-b75c-9406bf47779c. Accessed 12 June 2015

Mavhura E, Manyena SB, Collins AE, Manatsa D (2013) Indigenous knowledge, coping strategies and resilience to floods in Muzarabani, Zimbabwe. Int J Disaster Risk Reduc 5:38–48

Moser C (2011) A conceptual and operational framework for pro-poor asset adaptation to urban climate change. http://siteresources.worldbank.org/INTURBANDEVELOPMENT/Resources/336387-1256566800920/6505269-1268260567624/Moser.pdf. Accessed 25 Sept 2020

Moser C, Satterthwaite D (2008) Towards pro-poor adaptation to climate change in the urban centers of low- and middle-income countries. http://pubs.iied.org/pdfs/10564IIED.pdf. Accessed 25 Sept 2020

Moser C, Stein A (2011) Implementing urban participatory climate change adaptation appraisals: a methodological guideline. Environ Urban 23:463–485

Motsumi S, Magole L, Kgathi D (2012) Indigenous knowledge and land use policy: implications for livelihoods of flood recession farming communities in the Okavango Delta, Botswana. Phys Chem Earth A/B/C 50–52:185–195

Moyo M, Mvumi BM, Kunzekweguta M, Mazvimavi K, Craufurd P, Dorward P (2012) Farmer perception of climate change and variability in the semi-arid Zimbabwe in relation to climatology evidence. Afr Crop Sci J 20:371–333

National Bureau of Statistics (NBS) (2014) Delta state information. http://www.nigerianstat.gov.ng/information/details/Delta. Accessed 22 Dec 2015

Nelson G, Rosegrant MW, Koo J, Robertson R, Sulser T, Zhu T (2014) Climate change impact on agriculture and costs of adaptation. http://www.ifpri.org/sites/default/files/publications/pr21.pdf. Accessed 11 July 2015

Okem AE, Bracking S (2019) The poverty reduction co-benefits of climate change-related projects in eThekwini Municipality, South Africa. In: Cobbinah P, Addaney M (eds) The geography of climate change adaptation in urban Africa. Palgrave Macmillan, Cham. https://doi.org/10.1007/978-3-030-04873-0_10

Omohode R (2012) How massive flood swept away 50 Isoko communities, rendered thousands homeless. http://www.urhobotimes.com/individual_news.php?itemid=820. Accessed 6 Apr 2015

Perez C, Jones EM, Kristjanson P, Cramer L, Thornton PK, Förch W, Barahona C (2015) How resilient are farming households and communities to a changing climate in Africa? A gender-based perspective. Glob Environ Chang 34:95–107

Spear D, Selato JC, Mosime B, Nyamwanza AM (2019) Harnessing diverse knowledge and belief systems to adapt to climate change in semi-arid rural Africa. Clim Serv 14:31–36

United Nations Development Programme (UNDP) (2014) Delta State development performance: agricultural sector report, 1991–2014. http://www.undp.org/content/dam/nigeria/docs/IclusiveGrwth/UNDP_NG_DeltaState_Agric_2015.pdf. Accessed 15 Nov 2015

8

Rainfall Variability and Adaptation of Tomatoes Farmers in Santa: Northwest Region of Cameroon

Majoumo Christelle Malyse

Contents

Abstract

The Santa agrarian basin being one of the main market gardening basins in Cameroon and one of the producers of tomatoes in the country is vulnerable to the impact of rainfall variability. The spatiotemporal variability of rainfall through the annual, monthly, and daily fluctuations has greatly affected the market gardening sector in general and tomatoes production in particular. Thus, given rise to the research topic "Rainfall variability and adaptation of tomatoes farmers in Santa North west region of Cameroon," its principal objective is to contribute to better understanding of the recent changes occurring in tomatoes production and productivity in Santa. To attain this objective, a principal hypothesis was formulated that rainfall variability instead of unnatural conditions or human constraints justifies changes observed in tomatoes production in Santa and resulting adaptation strategies developed by peasants and stakeholders.

M. C. Malyse (✉)
Department of Geography, University of Dschang, Dschang, Cameroon
e-mail: majounachristelle@gmail.com

Our study came out with several findings, among which includes rainfall events in Santa fluctuate in time and in space with reduction in the number of rainy day and increase in the intensity of rainfall events causing soil erosion, infertility, and frequent crop diseases, insects, and pests. Extreme events such as drought and flooding have equally become frequent in the area especially during the different cycles of tomatoes production disrupting the agricultural calendar and causing crop failure and decrease in yields with Pearson's correlation of 0.017. This positive value shows that there is a relationship between annual rainfall and tomatoes output in Santa. Tomatoes farmers in Santa are struggling to adapt locally to this situations, but their efforts are still limited especially due to their low level of education and poverty. Finally, it was seen that the output of tomatoes over the years in Santa has a strong correlation with rainfall. Based on the findings of this study, the government is called upon to assist farmers in their adaptation options.

Keywords

Rainfall variability · Tomatoes production · Extreme events · Santa agrarian basin · Market gardening

Introduction

The change in global climate system is now undeniable, and it is human-induced as it has been concluded to a large extent by scientists during the past decade (IPCC 2007). Climate variability in general and rainfall variability in particular are very important environmental problems affecting mankind today with the highland areas being very sensitive to the change in rainfall pattern and its related impacts. Tsalefac (1999) in his work on climate variability in the western highlands of Cameroon brought out some aspects. He studied the relationship that exists between climate variability, land-use pattern, and the economic crisis that brings into question the sustainability of land use in the region. As such, this work explores the state of rainfall variability in Santa and some of the adaptation options adopted by tomatoes farmers in Santa. It's aim is to elucidate how changing rainfall pattern in space and in time as well as an increase in extreme weather events which includes dry spells, torrential rains, flooding, and droughts, which has greatly affected tomatoes yields, incidence of weeds, insects, pest and diseases, the economic cost of tomatoes production and how farmers are struggling to adapt to these changes?. This work analyzes tomatoes yields with production cost from 2001 to 2011 with efforts to understand how they have varied with the current rainfall changes. It focuses on extreme weather events on tomatoes production in Santa using examples from recent past such as severe drought of 1999 and 2004 that greatly disrupted the agricultural calendar. Finally, with the current rainfall and tomatoes scenarios serving as models, this work looks at better adaptation options that could limit the impact of rainfall variability on tomatoes production and boast tomatoes production in Santa.

The study of Zorom et al. (2013) and Rodriguez Solorzano (2014) on climate variability and adaptation practices by farmers shows the adoption of numerous adaptation strategies to cope with drought conditions. Some of the strategies adopted included the diversification of nonfarm activities such as selling of poultry and rearing of livestock as alternative measures as well as the reduction of food intake. Again, some farmers also engage in the cultivation of dry season irrigated vegetables. They further argued that these practices help farmers to distribute climate risk over different activities which strengthen their financial capacities to be able to raise their purchasing power.

These adaptation strategies adopted by tomatoes farmers in Santa range from changing production techniques to the adoption of improved tomatoes species, irrigation, and the use of agrochemicals and the diversification of activities. Stakeholders' adaptation ranges from the provision of agrochemicals to farmers finance to the provision of farm tools and farm equipment to farmers to help them improve on their production.

Tomatoes production which is a major market gardening crop in Santa has been witnessing a change in the daily, monthly, and annual rainfall in the area.

Variation in the date of onset, amounts, and the retreat of rainfall has affected the growth of tomatoes over the years.

However, the inability of tomatoes farmers to master and cope with rainfall especially the coming of rains has been a major setback to current and future tomatoes production as according to most of the peasants, rainy season during the past years has been very fluctuating and does not begin in mid-March as in the past years; rather, it begins in February and at times comes late and most of the time very is irregular, making the availability of water for irrigation very difficult. The few streams and watershed whose water up wells the water needed by plants are vast declining in quantity; this has been a natural occurrence, and these farmers and their crops have an average chance of survival. They loss income, time, and crops.

Dry spells as well as rainy days recorded on this area have been responsible for the outbreak of diseases such as septoria leaf spot, anthracnose, and *Verticillium* which leads to a fall in yields. This is particularly worrisome because farmers have very little strategies as they have resorted in the use of pesticides which has both negative and positive impacts on both the plant and the environment. Most of the time, pesticides are being washed away by heavy downpour due to the fact that most farmers do not know the best times to apply these pesticides. Heavy downpours have led to rapid depletion of the soils in some areas which have completely destroyed large portions of lands with crops. With all these, farmers in Santa have tried to make up for this loss by augmenting the dosage of fertilizers to beef up growth and productivity, but these fertilizers are very bad for the longevity of crops. Looking on the above explanations, the following question will guide us throughout this work: How does rainfall impact tomatoes farming in Santa? What are the farmers doing to remedy the situation?

Location of the Study and Research Methodology

Location of the Study Area

Santa is one of the 32 subdivisions in the northwest region located between latitude 5°42 N to 5°53 N of the equator and longitude 9°58 E to 10°18 E of the Greenwich meridian. It falls within the western highlands agroecological zone and covers a surface area of about 532,67 km^2. It covers some villages, namely, Akum, Baba II, Pinyin, Baligham, Matazem, and Santa (Njong, Ntarrah, and Mbei), which are our zone of study. This area lies some 20 km from Bamenda and is commonly called "the gateway into the northwest region" with an altitude from 1000 to 2600 m, making the area suitable for the cultivation of market garden crops especially tomatoes (Fig. 1).

Methodology

Data Collection

Primary data was collected from key informant such as tomatoes farmer, agricultural extension officers (MINADER and ACEFA), and field officers of research institutes (IRAD). This information was obtained through observations, interviews, and questionnaires. Field observation was very important in making a correlation of questionnaire response by the farmers to the actual remarkable activities in the farm. Field visits were made at different tomatoes farming sites in Santa where different farming, impacts of rainfall variability on tomatoes plants, and adaptation practice were applied with the help of some main tomatoes farmers. A total of 100 questionnaires were taken to the different selected villages, of which 39 were distributed in Ntarrah, 22 in Njong, and 39 in Mbei. Interviews were equally conducted with Farmers in the area who knew the area and had experience in tomatoes production. However, rainfall and output data was equally a secondary data of this study, of which the rainfall data here was obtained from two meteorological stations: Santa and Bamenda stations. The rainfall data used in the study ranged from 1963 to 2011 for the Bamenda station and from 1981 to 2006 for the Santa station. The different quantity of tomatoes harvested by farmers between the years 2001 and 2011 was obtained from the subdivisional delegation of agriculture for Santa and was used hand in globe with the rainfall data to show how rainfall variability has impacted on tomatoes production output over the years in Santa.

Presentation of Results and Discussions

The state of rainfall variability in Santa is analyzed based on the interannual and monthly anomalies as well as variability in rainfall intensity and rainy days which has really show evidence of rainfall variability in Santa this has greatly impacted

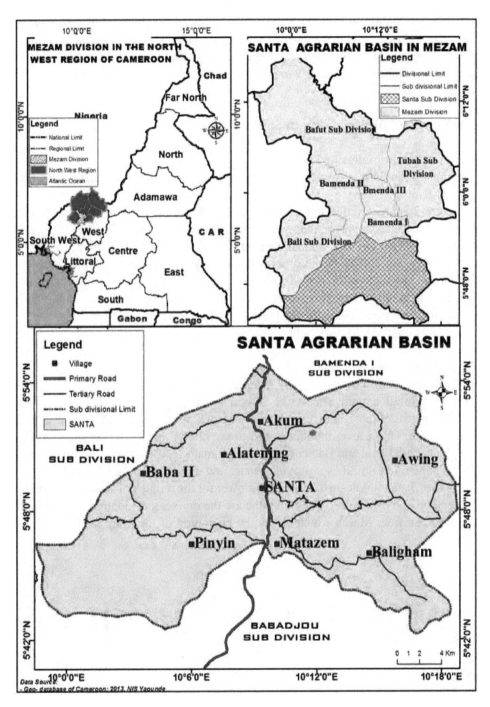

Fig. 1 Location and layout of the Santa agrarian Basin. (Source: Extracted from Fogwe and Zoum 2014)

tomatoes production in different ways which includes the attack of tomatoes crops by diseases, insects and pest, soil erosion causing infertility of soils, inaccessibility, water shortages most farmers have come out with several adaptation strategies to adapt to the impact of rainfall variability in Santa which has greatly decrease the yields of farmers.

State of Rainfall Variability in Santa

Interannual Anomalies in Santa
The situation in Santa demonstrates more negative anomalies than positive anomalies as seen.

Figure 2 shows rainfall anomalies in Santa. As already mentioned, years with positive anomalies are fewer than years with negative anomalies. The few years that registered rainfall over the annual average for the periods were 1990, 1992, 1997, 2001, 2002, and 2003. On the other hand, most years had rainfall lesser than the annual average such as the period from 1981 to 1989, 1993 to 1996, 1998 to 2000, and 2004 to 2006. Within these years of less rainfall, the period 2004 to 2006 stands out exceptionally and indicates dry periods.

Rainfall Variability in Santa
The variability of rainfall equally manifests on monthly basis. Generally, the monthly rainfall pattern affects the seasonal pattern. This area falls within the humid tropical climate with two distinct seasons, a short dry season and a long rainy season. Dry season months record lower rainfall amounts than rainy season months in both Santa and Bamenda stations. Tomatoes production in Santa depends on the pattern of rainfall as it provides water and moisture for plant growth.

Figures 3 and 4 indicate that rainfall is lower for the months of January, February, November, and December. These months are the dry season months. The amounts begin to rise from March which marks the beginning of the rainy season, and the

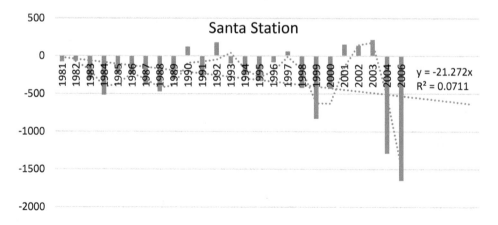

Fig. 2 Interannual anomalies

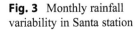

Fig. 3 Monthly rainfall
variability in Santa station

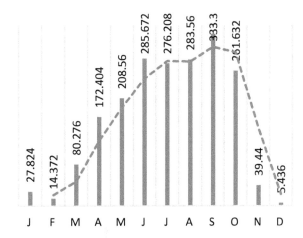

Fig. 4 Monthly rainfall
variability in Bamenda station

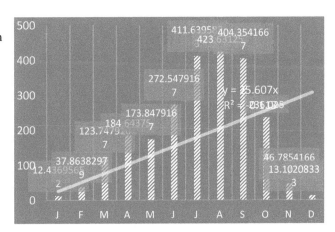

highest amounts are registered in the months of June, July, and August. These
months are at the mid of the rainy season.

This trend is the same for both stations, but there is an indication of spatial
variability when referring to the two stations. It is observed that rainfall amounts
vary within the same months in both stations.

Variability in the Number of Rainy Days and Rainfall Intensity

Variation in the number of rainy days and changes in rainfall intensity are some
powerful indicators of rainfall variability. This change has a link with fluctuations in
the dates of onset and departure of rains. Generally, a rainy day refers to a day that
records at least 1 mm of rainfall. The number of rainy days varies according to
seasons, but when observed annually, we noticed the days have been fluctuating
over the past years in Santa.

Figure 5 shows variability in rainy days from 1963 to 2011 in Santa. The highest
number of rainy days was recorded in 1999 (225 days), 1976 (222 days), and 2002

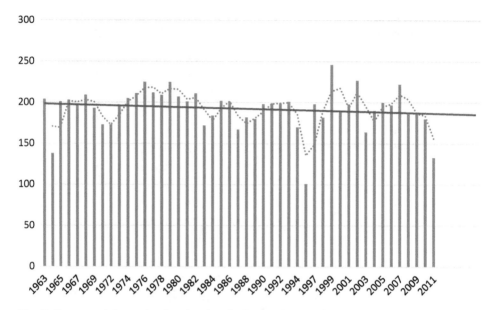

Fig. 5 Interannual fluctuation in rainy days in Santa. (Source: Fieldwork 2018)

Table 1 Farm-level adaptation strategies

	Variables	Frequency	Percentages
1	Crop diversification	17	17.34
2	Use of agrochemicals	54	55.10
3	Mixed cropping	4	4.08
4	Irrigation farming	13	13.26
5	Changing crop varieties	10	10.20
	Total	98	100

Source: Fieldwork 2018

(207 days). On the other hand, 1964, 1995, and 2011 had the least number of rainy days (138 days, 101 days, and 133 days, respectively). The annual difference in rainy days is about 124 days which is very large. This is an indicator of rainfall variability because it affects the frequency and intensity of rainfall.

Peasant Adaptation to Rainfall Variability on Tomatoes Production in Santa

Peasants in Santa have adopted several adaptation options to reduce the impact of rainfall variability on tomatoes production. These impacts are numerous, and the capacity to adopt varies according to farmers and the farm sizes. These adaptation options can be classified into two groups: the farm-level adaptation options and the nonfarm-level adaptation options (Table 1).

Tomatoes farmers in Santa use the farm-level adaptation options such as the use of new species of tomatoes, pesticides, and agrochemicals, changing farming

Photo 1 Improved variety of tomatoes (F-hybrid species). (Source: Photo by Majoumo 2018)

methods, and irrigation methods in order to reduce the impact of rainfall variability on their tomatoes farms.

Adaptation to Improved Varieties of Tomatoes

The use of improved varieties of tomatoes by most farmers in Santa has been a major response to drought and pest. The new variety used by farmers in this area is the F1-hybrid. According to one of the farmers interviewed (Photo 1):

> Hybrids are plants that are a result of cross pollination and they have advantages compared to open pollinated varieties they mature earlier and are uniformly than the other species and are resistant to change in seasons and diseases. (NGU 2018)

Adaptation Through Irrigation

Irrigation is one of the major adaptive strategies that could improve tomatoes yields in era of climate variability (Olwoch and Tshiala 2010). Tomatoes are not resistant to drought. Yields decrease considerably after the short period of water deficiency. It is important to water the plants regularly during flowering and fruit formation. The amount of water that is needed depends on the type of soils. Irrigation is one of the adaptive measures used by famers in Santa during periods of drought and low rainfall. Some of the different irrigation methods used by farmers in Santa on their farms includes the use of water pipes and watering canes to obtain water from wells to water farms, as can be seen in Plate 1.

From Plate 1, Photo 1 shows the sprinkler irrigation method used by tomatoes farmers in Santa to irrigate their farms during period of no rainfall or prolonged dry season, Photo 2 shows the water pipes used by farmers to connect water into their

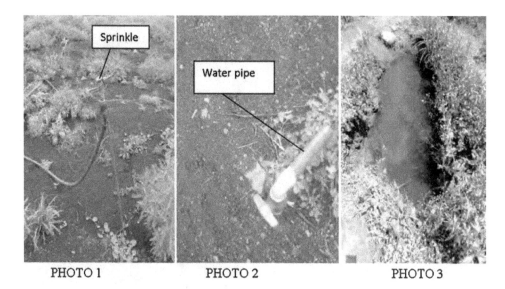

PHOTO 1   PHOTO 2   PHOTO 3

Plate 1 The different methods of irrigation used by tomatoes farmers in Santa

farms, and Photo 3 shows another local way of irrigation where farmers dig holes where there is water and use watering canes to carry water to water their farms.

Cropping System and Change in the Technique of Production as an Adaptation Option

Most farmers in Santa use poor farming methods to cultivate their crops which has greatly reduced their yields and also makes the area vulnerable to extreme climatic conditions such as flooding, droughts, as well as soil erosion. Tomatoes farmers in Santa are moving from monoculture to mixed cropping, crop rotation, and intercropping to avoid the risk of reduction in yields as well as crop failure and to equally allow the soils to regain its fertility. It was observed that tomatoes farmers in Santa mix tomatoes with crops such as green beans and lettuce and most of them do carry out intercropping on tomatoes farms.

Adaptation Through the Use of Agrochemicals and Fertilizers

The application of agrochemicals and fertilizers is common to all tomatoes farmers in Santa. The application of agrochemicals and fertilizers improves soil fertility and eradicates crop diseases, insects, and pests. Also, the use of agrochemicals especially fertilizer is a good adaptation strategy that improves tomatoes yields (Tshiala and Olwoch 2010). Most of the farmers in this area apply the NPK 20:10:10 fertilizer in order to increase fertility of their soils especially soils that have been rendered fertile during heavy rainfall or changing rainfall. With the growth of diseases and pest in Santa that affect tomatoes crops, most farmers adapt to use agrochemicals for their crops. An example of pesticide used by farmers in Santa is the CLEANZEB BLUE.

Table 2 Nonfarm-level adaptation

	Variables	Frequency	Percentages
1	Diversification of nonfarm activities	45	45.91
2	Migration	53	54.08
	Total	98	100

Source: Fieldwork 2018

Diversification to Off-Farm and Nonfarm-Level Activity Adaptation Strategies
Diversification to nonfarm and off-farm activities is one of the adaptive measures that is not popular to tomatoes farmers in Santa, but it was much preferred by most them as it was seen that most of the farmers in that area do carry out other activities such as trading, bike riding, and dress making, just to name a few. Some of these farmers even migrate to other areas in search of jobs especially in urban centers as shown in Table 2.

Table 2 shows the different off-farm adaptation options adopted by farmers in Santa. 54.08% of farmers say they diversify into nonfarm activities such as trading, teaching, and bike riding, just to name a few, while 45.91% migrate to other areas to obtain lands that are more fertile and to areas they think do not suffer from extreme weather conditions or in search for jobs. Respondent explained that the tomatoes business is the main source of livelihood bequeathed to them by their forefathers.

Conclusion and Future Prospects

This study has been able to come out with the state of rainfall variability on tomatoes production in Santa and some of the adaptation options adopted by farmers to remedy the situation. Rainfall events in Santa fluctuate seriously around the mean values. Data from two meteorological stations were analyzed and show yearly, monthly, and daily rainfall variability.

Spatial rainfall variability in the area shows that rainfall has become intense and fall over short duration varying from one area to another. This was analyzed in this work due to the fact that data for this study was obtained from two stations found on the same geographical locations. Temporally, as years goes by, rainfall is becoming very unreliable and varies greatly within the same months.

Extreme weather events such as heavy rains, drought, and flooding are increasingly becoming common in Santa, their occurrence determines the length of growing period, distorts the plants through diseases insects and pest during the plants life cycle, soil lost through flooding and erosion is one of the characteristics of rainy seasons in this area. Most of the farmers in the area are becoming aware of the situation as they have perceived the presence of rainfall variability in the area through the different ways in which it manifest.

In order to adapt to the situation, most of the farmers resorted in the use of irrigation, agrochemicals, mixed cropping, and diversification in nonfarm activities which have greatly helped them to improve on their productivity over the past years.

Future Prospects

The introduction of various adaptation strategies by the government and/or stake-holders, including the method of distribution of the agricultural inputs, was found to be very uncertain and partly unacceptable by most farmers in the study area. Thus, there is a need of further studies in this area to explore and develop the best method that will be efficient and effective.

Furthermore, an improvement in agricultural research will provide a backbone for adaptation measures. This is because research rapidly changing situations is different from research for stable conditions. Therefore, traditional knowledge is a suitable entry point but very insufficient in a changing situation. So tomatoes varieties need to be developed for future conditions as their applicability cannot be assessed at the location where they may be used in the future. The results of the research have to be published in an environment in which methods and crop varieties are accessible for use.

The analysis of the effect of rainfall variability on tomatoes production indicates that farmers do face production losses since the study area is vulnerable to rainfall and tomatoes production and tomatoes are one of the market gardening crops mostly cultivated by farmers in Santa. It is very important to develop a drought-resistant tomatoes variety that can withstand drought and extreme weather conditions and help to improve yields. It is therefore imperative that institutes such as crop research institutes like IRAD in the country develop more resistant drought varieties to withstand extreme weather conditions.

References

Fogwe ZN, Zoum BC (2014) Perception and adaptation ajustements to climate variability by Farmers within thé Santa agrarian Basin, 27–31pp.

IPCC (2007) The fouth assessment report

Olwoch JM, Freddy TM (2010) Thé impact of climate change and vraiblility on tomatoes production in thé limpopo province South Africa, 88–99pp

Rodriguez S, C. (2014) Unintended outcomes of farmers' adaptation to climate variability: deforestation and conservation in Calakmul and Maya biosphere reserves. Ecology and Society. Vol. 19, No 2 , 2pp

Tsalefac M (1999) Variabilité climatique, crise économique et dynamique des milieux agraire sur les Haute terres de l'ouest Cameroun. Thèse de Doctorat d'Etat de lettre et the science humaine, spécialité, option climatologie, Université de Yaoundé, 564p

Tshiala MF, Olwoch JM (2010) Impact of climate variability on tomato production in Limpopo province, South Africa. Afr J Agric RES. 13-20p

Zorom M, Barbier B, Mertz O, Servat E (2013) Diversification and adaptation strategies to climate variability: a farm typology for the Sahel. Agric Syst 116:7–15

9

Gender Implications of Farmers' Indigenous Climate Change Adaptation Strategies Along Agriculture Value Chain in Nigeria

Olanike F. Deji

Contents

Abstract

Climate change contributes significantly to the looming food insecurity in the rain-fed agricultural countries of Africa, including Nigeria. There is a gender dimension in climate change impacts and adaptation strategies along Agriculture

---

O. F. Deji (✉)
Obafemi Awolowo University, Ile Ife, Nigeria
e-mail: dejiolanike@yahoo.de

Value Chain (AVC) in Nigeria. The chapter gender analyzed the aspects of climate change impacts; identified the indigenous and expert-based artificial adaptation strategies; assessed the gender differences in the adaptation strategies; and provided the gender implications of the indigenous adaptation strategies among actors along the AVC. The chapter adopted a value chain-based exploratory design with gender analysis as the narrative framework with Gender Response Theory as the theoretical background. There were gender differences in the production, economic, and social dimensions of the climate change impacts along the AVC. The indigenous climate change adaptation strategies were availability, low cost, and easily accessible; hence they were popularly adopted by male and female AVC actors. The adopted indigenous adaptation strategies challenged the social relations, influenced reordering of social and gender relations, participation, and power relation among the male and female actors along the AVC.

Keywords

Gender · Farmers · Indigenous · Climate change · Adaptation strategies · Gender Response Theory · Agriculture Value Chain

---

Introduction

Food security influences agricultural production security. However, the diversion of government attention and political will from agriculture to the oil economy marked the beginning of food insecurity in Nigeria. While the population continues to increase, food production is decreasing at an alarming rate. Agriculture is a significant sector that impacts the socioeconomic livelihoods of the majority of the people of Nigeria, because it is the primary source of livelihoods for the majority. Unlike the oil booms that give direct benefits to a few minorities, Nigeria possesses a pro-agriculture environment and climate. Until the discovery of the oil boom in Nigeria, which marks the sharp diversion of the economic and political attention of the government away from agriculture to the oil boom, Nigeria was food sufficient and a significant exporter of most significant crops in Nigeria. The gap between the rich and the poor is widening as the benefits from oil booms increases, and agricultural production decreases.

Downie (2017) identified uncompetitive environment for agribusiness, inadequate input supply, poor market access, poor access to credit, lukewarm political commitment, and neglected agricultural research system as the obstacles to agricultural development in Nigeria. However, global warming-induced climate change and variability are worsening the situation. Ground-based observations and satellite data from the United States National Aeronautics and Space Administration (NASA) revealed that the first 6 months of 2016 were the warmest 6-month period since 1880 when records of temperatures begin (NASA 2016). The two major ice sheets are melting much faster relative to the past decades (Intergovernmental Panel on Climate

Change-IPCC 2014). During 2003–2013, disasters cost nearly US$1.5 trillion in global economic damage (Food and Agriculture Organization-FAO 2015).

The impacts of climate change and variability on agriculture is higher in the rain-fed agricultural nations such as Nigeria. Climate change is a significant push factor in agriculture in Nigeria due to the resultant irregular rainfall pattern and temperature swings resulting from climate change impacting agricultural production negatively along the value chain (Christopher and Jonathan 2011; Apata 2013; Acosta et al. 2015).

The impacts of climate change along the Agriculture Value Chain (AVC) have gender differences. Gender roles and gendered access and control over resources influence the gender difference in the vulnerability to climate change impacts by the male and female actors. The female gender is the most disadvantaged in terms of access to and control over resources, participation and contribution to decision-making at all levels, which makes them deficient in asset base; hence they are highly vulnerable to disasters such as from climate change and variability (FAO 2013; Gutierrez-Montes et al. 2018). This chapter explains the theory underpinning responses (adaptation/mitigation) of male and female AVC actors to the stimuli from climate change and variability.

Literatures (FAO 2011; Okali 2012; IPCC 2014; UNDP 2014) revealed that factors enhancing females' high vulnerability to the effects of climate change and variability are: (i) low adaptive capacity; (ii) fewer endowment and entitlements than men; (iii) unequal survival opportunities (for instance, limitations in mobility/migration); (iv) low decision-making potentials; and (v) inadequate access to and control over critical resources.

Kolawole et al. (2014) and Williams et al. (2019) affirm that the majority of farmers in Africa are smallholders and depend on local instead of scientific meteorology information for their adaptation to climate change and variability. Weather forecasting and early warning information to reduce vulnerability to climate change and variability are not readily available to farmers, are mostly written in technical jargons not smoothly comprehensive to local farmers, and are usually expensive to access (Raymond et al. 2010; Kolawole et al. 2014; Myuri et al. 2017). Scientific meteorology information requires technical skills to understand and adopt. Indigenous meteorology information is acquired through experiences and socialization by the parents and elders. According to Kolawole et al. (2014), both local and scientific meteorology information are products of observation, experimentation, and validation. However, scientific meteorologist adopts systematic procedure, while the process for the local one is unregulated, unorganized, and limited; hence the later may be less accurate and valid but could serve as stepping stone to the former. The local meteorology information could be useful in establishing a local weather experimental station to enhance farmers' access to the information for improved adaptation to climate change impacts.

Projects centered around the climate-smart agriculture (CSA) approach usually promote the adoption of technologies, as well as practices and services aimed at increasing agricultural productivity while enhancing producers' climate adaptation and mitigation capacities (Louman et al. 2015; Williams et al. 2019).

According to Gutierrez-Montes et al. (2018), if women have equal access to such technologies and practices and take ownership over the resulting benefits, Climate Smart Agriculture (CSA) may have a more significant effect on family well-being. To better understand the relationship between CSA, gender, and rural livelihoods, there is a need for well-defined and efficient indicators (SMART indicators) that allow project managers and policy-makers to assess and evaluate CSA programs or interventions in terms of their impact on gender relations (Gutierrez-Montes et al. 2018)

Women are known to be more involved in agricultural activities than men in sub-Saharan African (SSA) countries, Nigeria inclusive with as much as 73% involved in cash and food crops, arable and vegetable gardening, 16% in postharvest activities, and 15% in agroforestry (FAO and ECOWAS 2018; FAO 2019). The percentage of work done by women farmers far outweighs that of men, especially in Nigeria; they are major stakeholders for sustainable development (Faniyi et al. 2018, 2019; FAO 2019; National Bureau of Statistics-NBS 2016). There is gender role differentiation of immense dimension within African agriculture. Women make a significant contribution to food production and processing, but men seem to take more of the farm decisions and control the productive resources (Anaglo et al. 2013; Eger et al. 2018).

Nigeria accounts for nearly 20% of continental GDP and about 75% of the West Africa economy; despite this dominance, its exports to rest of Africa was at 12.7%, and only 3.7% of total trade is within the Economic Community of West African States (FAO and ECOWAS 2018; Aduwo et al. 2019). Despite the prominence of oil in the country's economic wealth, agriculture still contributes significantly to the Nigerian economy. The country's agriculture sector provides direct employment for about 75% of the population (NBS 2016; Alao et al. 2014). In the 1970s and 1980s, agriculture contributed nearly two-thirds of Nigeria's GDP. Currently, it provides about 40.2%, employs approximately 70% (males and females) of the labor force, accounts for more than 70% of non-oil exports and, most importantly, provides over 80% of the country's food needs (FAO 2010; FAO and ECOWAS 2018). With a population of over 180 million and still growing, agricultural development is vital for the attainment of food security and sustainable development in Nigeria.

Women farmers in rural areas are the majority of the agricultural workforce and should be empowered and provided free access to resources and participation in decision-making and programs (Bayeh 2016).

Women constitute a significant part of the agricultural labor force, and their contribution is essential to the success of the Economic Recovery and Growth Plan (ERGP) in the Federal Republic of Nigeria (FAO 2019; Faniyi et al. 2018). Although Women's roles are evident along the Agriculture Value Chain, the economic reward is not commensurate; they are not adequately benefiting from agricultural policies, programs, and budgets.

United Nations Development Program (UNDP 2014) report revealed that in Nigeria, women play a dominant role in agricultural production where they make up some 60–80% of the farm labor force, depending on the region, and they produce two-thirds of the food crops. Yet, the female farmers are among the voiceless, especially concerning influencing agricultural policies, programs, and development.

In Nigeria, a wide gender gap exists, and women in agriculture are worse for it (Aduwo et al. 2019). Nigeria agriculture is a rural community based, and 70% of the poor in Nigeria are in the rural areas, 59% of the poor household heads have women as heads (FAO 2011, 2013). Women constitute 70% of agricultural labor force; 60–70% of food crop producers; approximately 100% of food processors; 80% of food storage and transportation from farm gate to village market; 90% of hoeing and weeding work in farms; and 60% of harvesting and marketing services (Christopher and Jonathan 2011; Apata 2013). Despite the significant position of women in agriculture, men make major farm decisions and have access to land. Most women do not have the right to landed property, are denied access to credit and relevant capacity building opportunities, information, and participation (Anaglo et al. 2013; Eger et al. 2018).

The Objectives

Specifically, the chapter:

(i)   Gender analyzed the dimensions of climate change impacts
(ii) Identified the indigenous and expert-based artificial adaptation strategies
(iii) Assessed the gender differences in the adaptation strategies
(iv) Provided the gender implications of the indigenous adaptation strategies among actors along the Agriculture Value Chain (AVC)

The Theoretical Framework: Gender Response Theory (GRT)

Gender Response Theory – GRT propounded by Deji (2019) is the adopted theory in this chapter. Gender Response Theory states that males respond to stimuli (push or pull factor) by substitution while females respond by addition. And that male's response is usually more prompt than the female's, based on higher economic and social potentials.

Pull factors include positive and attractive forces such as new technology/innovation/idea/knowledge, to mention a few. Push factors include negative/repelling/adverse situations or circumstances such as climate change/variability, conflicts, poverty, ill-health, natural disaster, to mention a few.

GRT propound that usually male will respond to stimuli by substitution (replacing the old with the new). In contrast, females will often respond by addition (building on the existing local conventional or currently adopted knowledge/innovation/technology, to mention a few).

GRT was proven at four main levels of response, namely:

(i) Knowledge (indigenous and expert-based) – socialization, awareness, evaluation
(ii) Attitude (cognitive, affective) – interest, willingness

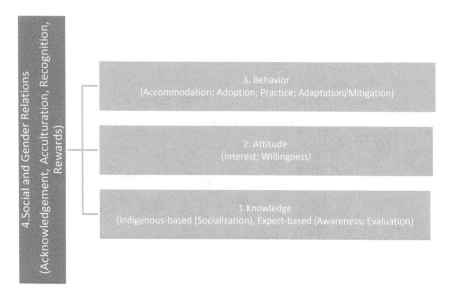

Fig. 1 Operational model of Gender Response Theory – GRT (Deji 2019)

(iii) Behavior (utilization of the knowledge as improvement or defensive/risk aver-
    sion strategy) – accommodation, adoption, practice, and adaptation/mitigation
(iv) Social and gender relations (responsibility, role, participation, engagement,
    decision-making, agency) – acknowledgment, acculturation, recognition,
    rewards (Fig. 1).

GRT confirms gender differences in the human response to stimuli (pull or push
factor). The theory affirms that: the indigenous knowledge is a common asset to both
male and female; males are likely to substitute their indigenous knowledge for new/
modern/scientific/expert-based knowledge, while the females are likely to add the
new knowledge to the indigenous knowledge; males are likely to respond to new
knowledge/innovation/technology more promptly than the females; and that the
economic and social relation potentials, especially the decision-making power,
influence the promptness of the male's response to (adoption of) the new knowl-
edge/innovation/technology, to mention a few.

Climate change is a push factor in agriculture, with significant impacts on rain-fed
agriculture that characterizes most developing nations like Nigeria. Male and female
are involved in activities along the AVC; climate change has implications along the
AVC and may have gender dimensions as pounded by the Gender Response Theory.
Logically, the male and female AVC actors will respond to climate change/variabil-
ity impacts more or less differently. Hence, GRT is the adopted theoretical framework for this chapter.

The following narrative sections in this chapter are textual, qualitative, and
secondary data, originated from field experience and literature through rigorously
digested knowledge and established information from the author's field experience
of over 20 years, covering all the ecological regions in Nigeria.

Agriculture Value Chain (AVC)

Agriculture Value Chain comprises interrelated activities and actors at different nodes from the point of decision and sourcing for inputs to the final stage when the agricultural product is processed, distributed, and consumed by the end users. The value chain in this chapter focused on crop cultivation (Fig. 2). Although the content of the activities may vary within different agricultural crop enterprises, they have many similarities.

1. Inputs: Includes finance, land, labor, tools and machines, seeds and seedlings and plant cuttings, chemicals, membership of agricultural associations and networks, extension and advisory services, training on required knowledge and skills, to mention a few. Required activities are sourcing, acquisition, transportation, storage, repair and maintenance, participation, to mention a few.
2. Land preparation: Includes activities such as land clearing, harrowing, landscaping, ridging, nursery bed construction, laying of irrigation pumps, to mention a few.
3. Planting: Planting/sowing, transplanting, grafting, budding, to mention a few.
4. Cultivation: Includes activities to enhance the germination and growth of the planted materials. It includes weeding, thinning, supplanting, staking, mulching, wetting and irrigation, fertilizer, and chemical applications, pest controls, transportation, to mention a few.

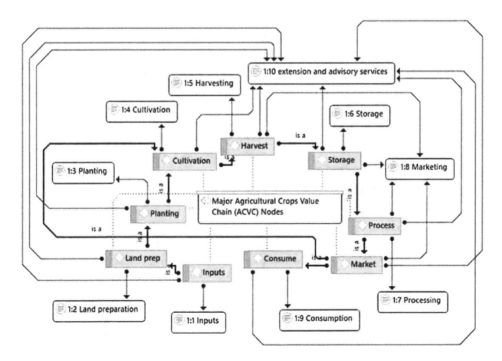

Fig. 2 Agriculture Value Chain (AVC) – Crops (Source: Deji 2019)

5. Harvesting: Harvesting, packing, transportation, marketing, consumption, to mention a few.
6. Storage: Packing, loading, transportation, application of preservatives, cleaning, and preservation, to mention a few.
7. Processing: Transportation,   conventional   processing,   value-addition, packaging, labeling, to mention a few.
8. Marketing: Transportation, packaging, selling, buying, record keeping, savings, advertisement, to mention a few.
9. Consumption: Purchasing, transportation, preparation, cooking, value addition, packaging, distribution, storage, to mention a few.

Gendered Dimensions of Climate Change Impacts Along AVC

Figure 3 shows the significant indicators of climate change and variability as experienced in Nigeria, such as irregular rainfall pattern and quantity; fluctuation in temperature; and increase in wind and storm intensity. Furthermore, Fig. 1 indicates the primary and secondary impacts of climate change and variability. The significant consequences include decrease in rainfall/water shortage; irregular temperature fluctuation; dry weather; decline in soil moisture and fertility; an increase in drought incidence; and a rise in flood incidence.

The secondary impacts have two dimensions along the AVC, such as (1) production and economical and (2) social and gender relations. The dimension of production and economical includes inappropriate planting periods; low and poor yields; loss of species; decrease in agro-biodiversity; increase in epidemics; and increase in postharvest losses. The social and gender relation impacts dimension encompasses the reordering in social and gender roles and relations along the AVC.

Figure 4 shows the significant impacts of climate change indicators such as irregular rainfall pattern and quantity, flooding, rise in temperature, and wind storms at each of the seven nodes along the AVC.

1. Input node (access and utilization): Climate change impacts are: scarcity of some inputs such as fertilizer due to the high rate of damages done by heavy rainfall and high temperature. Poor quality of some inputs such as seedlings due to temperature and rainfall fluctuations; increase in the cost of production arising from spending more money and time; and high rate of damages/breakdown and difficulty in using farm machinery.
2. Land preparation: Major climate change impacts include unfavorable soil moisture and texture; loss of topsoil; increase in frequency of land preparation activities such as plowing and harrowing due to an increase in the rate of weed growths; and increase in cost.
3. Planting: Unpredictable planting time resulting to early and delay planting; repeated planting and thinning; utilization of more planting materials and resources like time, money, and labor; increase in wastages of resources like time, money, and labor; and higher pressure on the land leading to soil hardening and loss of fertility.

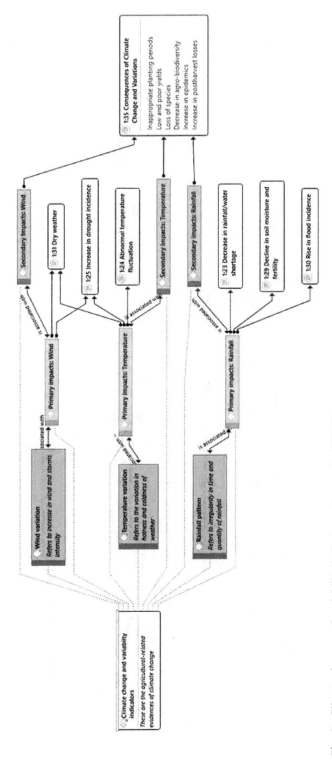

Fig. 3 Climate change and variability indicators. (Source: Deji 2019)

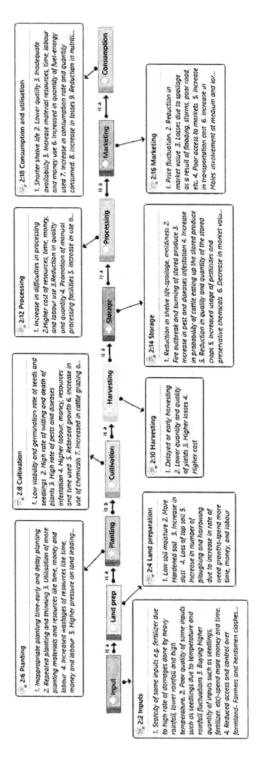

Fig. 4 Climate change impacts along Agriculture Value Chain (Source: Deji 2019)

4. Cultivation: Climate change influences low viability and germination rate of seeds and seedlings; plant growth retardation; high rate of wilting and death of plants; high rate of pests and diseases infestation; higher cost in forms of labor, money, resources, and time used; increase in use of chemicals; and increase in cattle grazing on the crops/farmland.

5. Harvesting: Climate change influences delayed or early harvesting, lower quantity and quality of yields; higher losses; and higher cost of harvesting.

6. Storage: Climate change significantly contributes to the reduction in shelf life of farm produce. It contributes to increased fire incidence and burning of stored produce, increase in pest and disease infestation, increase in the probability of cattle eating up the stored produce, and reduction in quality and quantity of the stored farm produce. It enhances the usage of pesticides and preservative chemicals; and decrease in nutrient and market value.

7. Processing: There is an increase in difficulties in processing farm produce; higher cost; reduction in quality and quantity of finished farm products; and promotion of manual processing facilities. It increases the use of preservatives and other chemicals.

8. Marketing: There is an increase in price fluctuation; reduction in market value; loss; poor access to markets; increase in transportation and other costs; increase in males' involvement at medium and large-scale marketing.

9. Consumption and utilization: At the household level, climate change influences shorter shelf life. It enhances low quality, inadequate availability of food materials; increase in cost of food preparation and readiness; increase in consumption rate and quantity consumed; increase in losses; and reduction in nutritional benefits, meal/food security.

Factors Influencing the Vulnerability of Agriculture in Nigeria to Climate Change Impacts

(i) Rain-fed agriculture
(ii) Subsistence farming
(iii) Gendered division of labor and access to resources
(iv) Lack or inadequate social insurance for farmers
(v) Unfavorable gender norms against women who are the major workforce
(vi) Patriarchal culture: practiced as males' domain
(vii) No formal insurance plan for farmers
(viii) Small scale and low literate by majority
(ix) Low awareness of the agriculture-based sources of greenhouse gases (GHG) among the farmers
(x) Limited modern effective adaptive and mitigation capability
(xi) Limited access to scientific meteorological information and technologies
(xii) Inadequate political will and enabling environment for climate smart agriculture

Gendered Indigenous Climate Change Adaptation Strategies Along Agriculture Value Chain

Climate change and variability impact are not gender neutral; they affect both males and females differently. There are complex and dynamic links between climate change and gender in terms of vulnerability to the adverse impacts of climate change and adaptation to climate change. Within countries, vulnerability to climate change impacts links to poverty and economic marginalization (UNDP 2014).

Adaptation is a process by which individuals, families, communities, and countries minimize the negative impacts of climate change and variability on their socioeconomic livelihoods. It means the strategies or methods adopted in coping with the threats resulting from the unavoidable climate change and variability impacts. On the other hand, mitigation means actions, practices, steps taking or adopted to reduce the greenhouse gases (GHG) in order to minimize their effects on global warming.

Indigenous adaptation strategies are the traditional conservative knowledge, experience, and practices that are products of repeated activities, communicated from parents and elders to younger ones through the socialization process, adopted as coping strategies to reduce the vulnerability and impacts of climate change. Indigenous adaptation strategies for alleviating the climate change impacts on agriculture and AVC actors' livelihood varied basically on the gender roles (direct and indirect) along the AVC. Indigenous adaptation and mitigation strategies are products of the repeated process of informal *observation, trial, experimentation,* and *validation,* which naturally promotes/popularizes the indigenous knowledge, experiences, and practices over time.

Roles are responsibilities carried out or performed by the primary or secondary owner of the responsibility. Gender roles are activities carried out or performed by an individual based on his/her sex as constructed by the norms and values of the society.

The concept of gender roles is significant and most applicable in agriculture discipline, because it helps to understand that some activities are carried out by individuals outside the assigned responsibilities as constructed in the norms and values of the society (Figs. 4a–e). Direct gender role refers to the activity personally carried out by an individual based on his/her sex as determined by the norms and values of the society. Indirect gender roles are activities carried out by a male or female (paid or free of charge) on behalf of another person, which may not necessarily be according to the norms and values about gender roles in the society.

Climate change significantly influences gender roles along the AVC (Figs. 5a–e). As a response to the negative impacts of the climate change on agriculture productivity along the value chain and the general livelihoods of the farm household, there are observed dynamics of roles between the male and female AVC actors. Such gender role dynamics include role delegation, role diversification, role commodification, and role intensification.

The dominant indigenous adaptation strategies among the AVC actors in Nigeria are as follows:

1. **Preproduction phase:** Bulk purchase of inputs at the group level, bush fallowing, zero tillage, fragmented planting, late and early planting, and land intensification
2. **Production phase:** Mulching, thinning, supplanting, manual irrigation, nursery and transplanting, organic fertilizer and micro-dozing application, staking, use of local herbs, mixed cropping, multi-cropping, crop rotation, mixed farming, dry season farming, traditional greenhouse farming, crop diversification, varying planting dates, increase in irrigation, soil and water conservation techniques, shading and shelter, and shortening the length of the growing season
3. **Harvesting phase:** Early/late harvesting, fragmented/installment/selected harvesting, on-farm sales of fresh crops, harvesting fresh
4. **Storage:** Air drying, sun drying, heat drying, e.g., over kitchen roofs, application of pepper, storage under the roof, over the kitchen, in gourds, earthen pots, storing below room temperature overnight on the roof, to mention a few
5. **Processing:** Value addition, repackaging to enhance economic value, group purchase of materials
6. **Marketing:** Online marketing, e.g., mobile phone, group-marketing of produce, value-addition, selling on the farm and at the farm gate, direct sales to bigger companies and consumers
7. **Consumption:** Avoiding bulk purchase, cooking what can is consumed at once, storage of used water for other purposes

The Indigenous and Expert-Based Artificial Adaptation Strategies

Male and female farmers and other value chain actors acquired indigenous adaptation and mitigation intelligence/strategies through experiences and socialization. The indigenous intelligence/strategies are popular local knowledge asset bases among the African AVC actors because they are available, cheap, and simple to understand and practice with limited gender discrimination. The indigenous adaptation intelligence/strategies are products of informal, unregulated observations, experimentations, and validation.

The indigenous knowledge/intelligence/strategy is different from the expert-based human intelligence (e.g., through extension agency) and the artificial intelligence (human-embedded machine intelligence). The indigenous knowledge/intelligence/strategy is not regulated, lacks accuracy, hence not usually efficient and effective.

There was low awareness and adoption of expert-based/artificial intelligence-based climate-smart agricultural/adaptation strategies. It may be due to their inherent low cultural compatibility, high cost, inadequate availability, gender limitation, and complexity that require high technicality and competence.

The indigenous adaptation strategies have the potentials to provide an enabling environment that could enhance the popularity and adoption of artificial intelligence

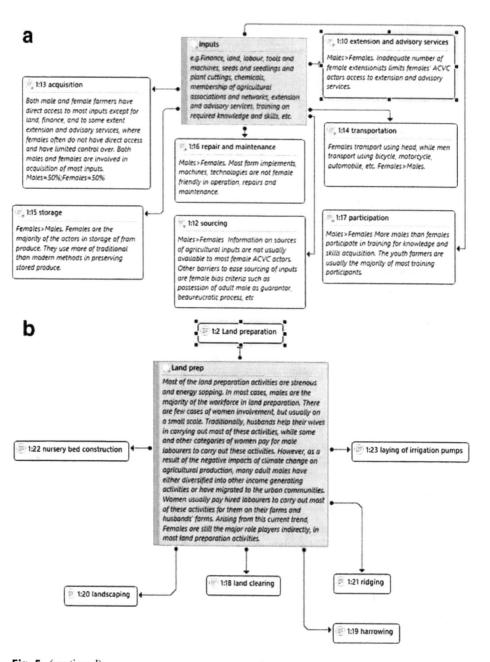

Fig. 5 (continued)

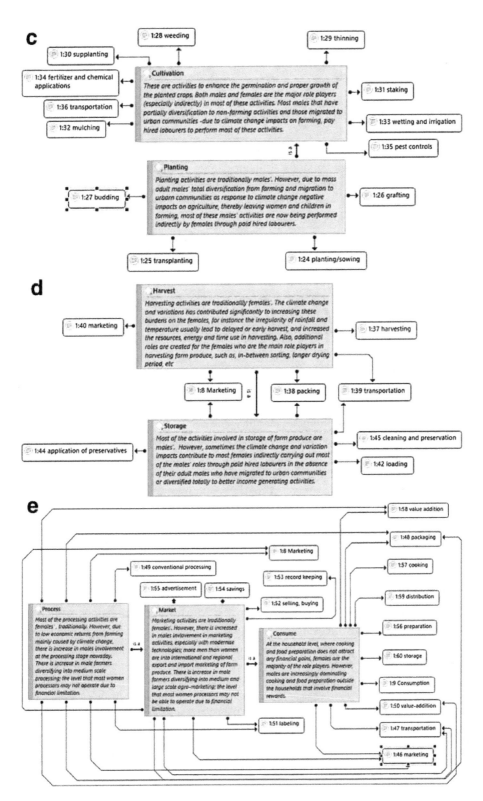

Fig. 5 (continued)

(AI) if adequately integrated. However, adequate integration will require an inter-disciplinary scientific reinvention process that engaged both male and female AVC actors and the extension professionals (expert-based human intelligence).

Gendered Indigenous Adaptation Strategies to Climate Change/ Variability

The popular indigenous adaptation strategies among Nigerian male and female AVC actors are in two categories: (i) resource-based indigenous adaptation strategies and (ii) relation-based indigenous adaptation strategies. Resource-based indigenous adaptation strategies target the factors of production-related adaptation strategies, such as land, finance, labor, and time (Fig. 5). The relation-based adaptation strategies focus on social capital and relations. The resource-based and relation-based indigenous adaptation strategies are:

1. **Land**: *Land fragmentation* for *multi-cropping* and *mixed farming*; *land commodification*: sales of farmland for money; *land intensification*: repeated cultivation on farmland leading to *shorter fallow period*; intensive practice of *inorganic soil fertilization and conservation* methods; and *land use diversification*: using the farmland for nonagricultural income-generating activities.
2. **Finance:** *Financial diversification*: investing financial resources in other income-generating activities such as farming and nonfarming activities; *multiple sources of finance*: to increase financial stability and security; increase in *contract farming*: an organization supplied farmers all essential inputs to produce certain crops that will be purchased in kinds by the organization; and *community-based cooperative or group saving, credit, and loan system*.
3. **Labor:** *Migration* of nondisabled males (spatially migrating from farming communities to urban cities in search of alternative sources of income, or livelihood-wise migration to other jobs, especially, nonagricultural income activities); *use of family and child labor*; *livelihood diversification* to other activities along AVC (for example, a crop farmer diversifying into retailing/marketing of farm produce) or to nonfarming activities; and increase in use of *hired labor*, especially to carry out strenuous and abandoned agricultural activities.
4. **Soil conservation management practices:** Most of the indigenous soil conservation management measures are both adaptive and mitigating in nature, e.g.,

Fig. 5 (**a**) Climate change impacts on gender roles along Agricultural Value Chain – input node. (Source: Deji 2019). (**b**) Climate change impacts on gender roles along Agricultural Value Chain – land preparation node. (Source: Deji 2019). (**c**) Climate change impacts on gender roles along Agricultural Value Chain – planting and cultivation nodes. (Source: Deji 2019). (**d**) Climate change impacts on gender roles along Agricultural Value Chain – harvesting and storage nodes. (Source: Deji 2019). (**e**) Climate change impacts on gender roles along Agricultural Value Chain – processing, marketing, and consumption nodes. (Source: Deji 2019)

agroforestry. *The fallow system* encourages forest development; the forest is a carbon sink. *Agroforestry* is effective in carbon sequestration; it is a rational land use planning system that tries to establish a balance in food crop cultivation and forestry, leading to an increase in organic matter in the soil, which indirectly reduces the pressure exerted on forests. *Bush fallowing*: the use of natural fallows – leaving the land uncultivated for more than 2 years to regenerate or restore soil fertility to regenerate or restore soil fertility. *Organic manure application*: the application of compost, animal waste, and domestic wastes to the soil to maintain soil microbial activities and promote absorption of nutrients by plants. *Intercropping*: cultivation of more than one type of crop on a piece of land at the same time to reduce the risk of total crop loss/failure as well as providing good soil cover that minimizes soil erosion. *Conservation tillage*: minimal or no disturbance of soil; minimum or zero tillage; to respond to rapid soil deterioration and degradation caused by conventional tillage under harsh weather induced by climate change and variation.

5. **Water conservation/management practices:** *Rain harvesting* in pits and wells, barrels, to mention a few; *reuse of used water*; *rehabilitating degraded land*, planting of *drought-resistant crops* such as cassava, sweet potatoes, indigenous finger millet, to mention a few.

6. **Crop and farm management measures (risk-aversion practices):** Most farmers in developing countries like Nigeria do not enjoy formal insurance for risk aversion on their farms. Instead, they employ indigenous crop and farm management measures as climate change adaptation strategies, including *fragmented planting and harvesting; mixed cropping; multi-cropping; multiple investments; group farming; contract farming; decentralized or dispersed farming* – farming in one than one location at a time.

7. **Social capital/relation conservation measures:** Along the AVC, most male actors adapt to the unfavorable economic consequences of the climate change impacts through *role delegation; role commodification; role diversification; child commodification/child labor; rural-urban migration*: total or partial, economic or spatial, dual residency; and *livelihood diversification* to nonagricultural-related activities (vertical). *Role delegation* is a practice whereby most men delegate (substitution) their farming and family roles to the females, leading to extra burdens (addition) on the females' time, energy, and resources (Gender Response Theory). *Role commodification* means the strategy whereby most male AVC that have partially migrated or diversified to other jobs pays someone else to carry out their agricultural roles on their farms. *Child commodification* refers to the practice of exchanging children for money or material resources.

   Horizontal role dynamics (by substitution) is popular among the males, while *vertical role dynamics* (by addition) is predominant among the female AVC actors. However, in livelihood diversification, male AVC actors diversify more into nonagricultural-related activities (*vertical livelihood diversification*). In contrast, the female AVC actors diversify mostly into agricultural-related activities (*horizontal livelihood diversification*) (Fig. 6).

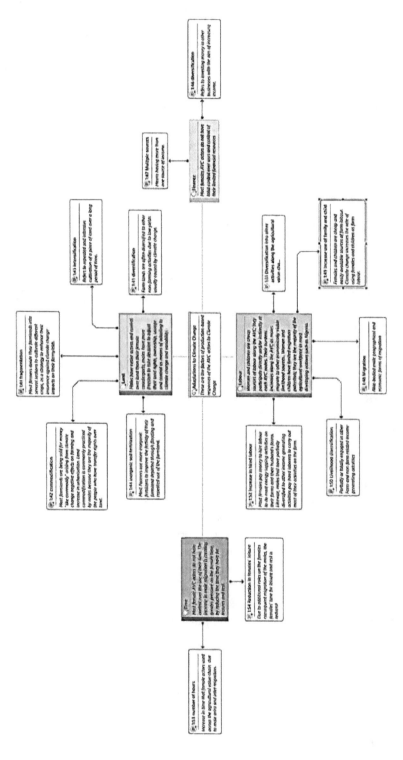

Fig. 6 Gendered resource-based adaptation strategies to climate change impacts. (Source: Deji 2019)

Indigenous Early Warning Meteorological Adaptation Strategies

Most farmers in Nigeria do not have adequate access to scientific meteorological information; they often depend on their indigenous early warning strategies. In the face of growing climate change that makes accurate prediction of weather and weather forecasting very difficult, most farmers continue with their indigenous early warning knowledge as adaptation strategies to the climate change impacts.

The indigenous meteorology practices on early warning measures and knowledge are products of years of farming experience and acquired knowledge from parents and elders in the farming communities (Table 1). Most scientific meteorology knowledge is costly and requires technical skills to understand and adopt. On the contrary, indigenous meteorology practices are readily available and straightforward to understand and adopt by an average African farmer characterized by low literacy and socioeconomic power.

Gender Implications of the Indigenous Climate Change Adaptation Strategies

Climate change is a significant threat to agricultural productivity and profitability, especially in a rain-fed agricultural nation like Nigeria. There is an increase in poor

Table 1 Some common African indigenous early warnings and weather prediction. (Source: Deji 2019)

Indigenous early warning	Weather prediction
1. General yellowing and falling of tree leaves	Onset of Harmattan season (winter)
2. Flowering of the peach tree (*Prunus persica*)	It is time to plant
3. Blooming of pear flower (*Pryus communis*) *or* roses (*Rosa damascene*)	Time to plow
4. Regular chanting of red-chested birds (*Cuculus solitaries*)	Time to plant regardless of the season
5. Half-moon facing east/full moon/rainbow on the sky	No rain to come
6. Increase in pests	More rain to fall
7.Increase in worms in the subsoil/surface of the soil	Season of summer and excessive rain/high water table/flooding possibility
8. Increase in frogs	Onset of thunder storms
9. Fruiting of brandy bush tree (*Grewia flava*):	
(a) Fruiting between November and December before the first rain in the year	Scanty or low rainfall
(b) Fruiting between February and March	Abundant rainfall
(c) No fruiting throughout the year	Drought
10. Flowering and fruiting of shepherd tree (*Boscia albitrunca*):	
Flowering and fruiting before the first rain	Year of abundant rainfall

agricultural yield leading to social and financial insecurity, reduction in farm-based income, and economic instability of the farm households. Indigenous adaptation strategies are commonly used among the AVC actors to reduce the negative impacts of climate change on their livelihoods. However, most of the indigenous adaptation strategies have implications on the social and gender relations, especially at the household levels, including:

1. Role: There is an increase in adult males' migration and diversification to nonagricultural income activities; female actors are more prominent at all the nodes along the AVC. Along the AVC, there is the high mobility of males from the production phase to the marketing phase, which promised better economic gains.
2. Reordering of social and gender relations: The male actors' spatial (rural-urban) and livelihood (from agriculture to nonagriculture) migration trend initiated by climate change is creating reordering of social and gender relationships within the farming households. Female AVC actors are increasingly occupying the spaces left along the AVC by most migrated males. Hence, there is an increase in females' involvement in decision-making, especially at the household level. Likewise, there is an improvement in females' access and control of resources at the household level.
3. Responsibility: There is an increase in responsibilities of and burdens on the female AVC actors. There is an increase in child labor and commodification (both males and females). Women and girls wake up early to fetch water for food preparation and irrigation of crops before herds come around the water to satisfy their increasing thirst due to an increase in temperature caused by climate change (which often results in herdsmen/farmers' conflict). Consequently, there is a reduction in the resting period for the females, which, in the long run, has negative health implications.

Conclusions

Climate change has the potentials to increase the vulnerability of the poor. The unfortunate situation and conditions of most agricultural value actors, especially the females, exposes them to negative phenomena and make them hard-bitten. Climate change impacts are felt more on the poor living in the rain-fed regions such as Nigeria. The chapter confirms that:

1. Agriculture in Nigeria is rain-fed in nature; hence it is significantly impacted negatively by climate change and variability.
2. Climate change is threatening the livelihoods of Nigeria Agriculture Value Chain (AVC) actors, the majority of whom are the female small-scale holders with limited access to scientific expert-based adaptation and mitigation practices, hence increased food insecurity and poverty.

3. The female AVC actors are mostly affected by the climate change impacts due to their domestic and cultural roles, limited migration potentials, and the discriminating gender norms that limited their easy access and control over production resources and decision-making.

4. More of the female than male AVC actors experience difficulty in accessing resources due to gender discrimination and gender irresponsive policies and criteria dictated by the traditional and legal norms in the society.

5. Climate change and variability are influencing the reordering of social and gender relations among the AVC actors, especially at the household levels, most of which have implications for women empowerment and gender equality.

6. The indigenous adaptation strategies are different from the expert-based human intelligence (provided through extension service) and the artificial intelligence (machine intelligence); the indigenous intelligence is not regulated, lack accuracy, and hence not efficient and effective.

7. The low awareness and adoption of artificial intelligence-based climate-smart agricultural/adaptation strategies among the AVC actors were due to their low cultural compatibility, high cost, inadequate availability, gender limitation, complexity that requires high technicality, and competence.

8. Both male and female AVC actors adopted indigenous adaptation strategies because they were culturally compatible, available, low cost, and simple to understand and apply, while expert-based artificial adaptation strategies were not common among the AVC actors.

9. Mobility, especially rural-urban, was a typical indigenous adaptation strategy among the male AVC actors. Females have limited mobility/migration potentials being the dominant workforce in the domestic and caregiving services, especially at the household level.

10. Role commodification, role diversification, role delegation, and land commodification were popular strategies among males.

11. Horizontal role dynamics (by substitution) were most common among the male AVC actors, while the vertical role dynamics (by addition) were most common among the female AVC actors as indigenous adaptation strategies to climate change. In response to climate change and variability impacts, there was significant role-based social mobility (vertical and horizontal) within each AVC node among the male and female value chain actors.

12. Most male AVC actors adapted to climate change and variability by substituting activities and roles along the AVC and diversifying, mostly, into nonagriculture livelihood sources. On the contrary, most female AVC actors adapted to climate change and variability by taking up additional activities and roles (role intensification), usually within the agricultural value chain, as confirmed by the "Gender Response Theory – GRT."

13. According to Gender Response Theory, indigenous knowledge/ strategies could provide an enabling environment for the sustainable adoption of modern artificial knowledge and strategies.

14. The theory further reveals that males are more likely to substitute new knowledge/strategies for their indigenous adaptation knowledge/strategy, and adapt

faster than the females due to their higher economic and decision-making potentials. However, a response by addition (females' response) is usually more sustainable than a response by substitution (males' response).

15. Climate change is challenging social relations, especially at the household and community level. Examples are an increase in conflict between herdsmen and farmers, land-based conflicts, and child commodification.

16. Climate change is challenging the gender relations; female AVC actors are increasingly prominent and active in most activities along the AVC and decision-making and agency at the household and community levels due to the prevailing migration and livelihood diversification among active-aged males.

Recommendations

Intervention programs and expert-based/artificial intelligence technology for climate change adaptation that will effectively enhance food security must recognize the indigenous strategies, knowledge, experiences, and practices, of the male and female AVC actors. Such programs must have in-built potentials to improve the social and economic situation and conditions as well as reduce the stress and burdens of the female actors along the AVC.

Furthermore, the male or female gender group is not homogenous on the bases of social variables such as age, education, ethnicity, social and economic status, marital status, family size, to mention a few. By implication, there might be variations in climate change impacts within each gender category at the different nodes along the AVC. Hence, there is a need for further gender analysis in AVC research using the intersectionality concept and methodology to enhance the inclusive understanding of the impacts and adaptation strategies within each gender group.

Also, research on climate change/variability must adopt an interdisciplinary approach using gender analysis methodology and theories for an inclusive under-standing of the experiences and adaptation strategies among the different categories of the male and female AVC actors.

Acknowledgments The financial supports of the **Alexander von Humboldt Foundation** and the material and moral supports from the **Friedensau Adventist University** and **Obafemi Awolowo University** are acknowledged.

References

Acosta M, Ampaire E, Okolo W, Twyman J (2015) Gender and climate change in Uganda: effects of policy and institutional frameworks. CCAFS Info Note. Copenhagen, Denmark. CGIAR Research Program on Climate Change, Agriculture, and Food Security (CCAFS). Available online at: http://www.ccafs.cgiar.org. Accessed 10 Dec 2019

Aduwo OE, Aransiola JO, Ikuteyijo LO, Alao OT, Deji OF, Ayinde JO, Oyedele DJ (2019) Gender differences in agricultural technology adoption in developing countries: a systematic review. Acta Hortic 1238. https://doi.org/10.17660/ActaHortic.2019.1238.24

Alao OT, Adebooye OC, Deji OF, Idris-Adeniyi KM, Agbola O, Busari AO (2014) Analysis of the impact of production technology and gender on under-utilized indigenous vegetables production in south-western Nigeria. Afr J Sci Technol Innov Dev 6(1). https://doi.org/10.1080/20421338.2014.931741

Anaglo JN, Boateng SD, Boateng CA (2013) Gender access to agricultural resources by small-holder farmers in the Upper West Region of Ghana. J Educ Pract 5(5):13–19

Apata TG (2013) Analysis of cassava value chain in Nigeria: from a pro-poor and gender perspective of farming households in southwest Nigeria. Merit Res J Agric Sci Soil Sci 2(11):147–153

Bayeh E (2016) The role of empowering women and achieving gender equality to the sustainable development of Ethiopia. Pac Sci Rev B Humanit Soc Sci. https://doi.org/10.1016/j.psrb.2016.09.013

Christopher C, Jonathan M (2011) Gender and agricultural value chains: a review of current knowledge and practice and their policy implications. ESA working paper no. 11–05

Deji OF (2019) Gender Response Theory (GRT); in climate change impacts on gender roles dynamics along agriculture value chain and implications for women farmers' empowerment in Nigeria: Why is the Intersectionality? A paper presented at Autumn conference of the Developmental Sociology/Social Anthropology Section (ESSA), Friedensau Adventist University, Germany, November 7–8, 2019. 48pp. And in climate change Adaptation and Mitigation strategies among farmers in Africa: Indigenous Intelligence vs Artificial Intelligence. A public lecture presented at Aula Hall, Friedensau Adventist University, Germany, December 11, 2019. 28pp

Downie R (2017) Growing the agricultural sector in Nigeria. A report of the Center for Strategic and International Studies (CSIS) global food security project

Eger C, Miller G, Scarles C (2018) Gender and capacity building: a multi-layered study of empowerment. World Dev. https://doi.org/10.1016/j.worlddev.2018.01.024

Faniyi EO, Deji OF, Alabi DL, Ijigbade JO (2018) Soil and water conservation capabilities of male and female vegetable farmers on micro-veg project sites in southwestern Nigeria. J Agric Ext 22(3). https://doi.org/10.4314/jae.v22i3.12

Faniyi EO, Deji OF, Oyedele DJ, Adebooye OC (2019) Gender assessment of vegetable farmers' utilization of soil and water conservation technologies in MicroVeg project sites, Southwest Nigeria. Acta Hortic 1238:239–248. https://doi.org/10.17660/ActaHortic.2019.1238.25

Food and Agriculture Organization of the United Nations (2010) Gender dimensions of agricultural and rural employment: Differentiated pathways out of poverty Status, trends and gap. Available: http://www.fao.org/docrep/013/i1638e/i1638e.pdf

FAO (2019) FAO and the 2030 agenda for sustainable development. https://doi.org/10.18356/ea058ce2-en

FAO and ECOWAS Commission (2018) National gender profile of agriculture and rural liveli-hoods–Nigeria. Country Gender Assessment Series, Abuja, 92pp

Food and Agriculture Organization of the United Nations -FAO, UN (2015) Review of climate change adaptation and mitigation in agriculture in the United Republic of Tanzania, mitigation of climate change in agriculture (MICCA) program. The impact of disasters on agriculture and food security. Food and Agriculture Organization of the United Nations, Rome

Food and Agriculture Organization (FAO) (2013) A tool for gender sensitive agriculture and rural development policy and program formulation: guidelines for Ministries of Agriculture and FAO. Food and Agriculture Organization of the United Nations, Rome

Food and Agriculture Organization of the United Nations (FAO) (2011) The state of food and agriculture: women in agriculture closing the gender gap for developing. Rome, 160p. http://www.fao.org/docrep/013/i2050e/i2050e.pdf

Gutierrez-Montes I, Arguedas M, Ramirez-Aguero F, Mercado L, Sellare J (2018) Contributing to the construction of a framework for improved gender integration into climate-smart agriculture

projects monitoring and evaluation: MAP-Norway experience. Clim Chang:1–14. https://doi.
org/10.1007/s10584-018-2231-1

Intergovernmental Panel on Climate Change –IPCC (2014) Intergovernmental Panel on Climate
Change (IPCC), 'Summary for Policymakers'. In: Field CB, Barros VR et al (eds) Climate
change 2014: impacts, adaptation, and vulnerability. Part A: global and sectoral aspects,
Contribution of Working Group II to the fifth assessment report of the Intergovernmental
Panel on Climate Change. Cambridge University Press, Cambridge UK/New York, pp 1–32

Kolawole OD, Wolski P, Ngwenya B, Mmopelwa G (2014) Ethno-meteorology and Scientific
weather forecasting: small farmers' and Scientists' perspectives on Climate variability in the
Okevango Delta, Bostwana. Clim Risk Manag Elsevier 4–5:43–58

Louman B, Campos-Arce JJ, Mercado L, Imbach P, Bouroncle C, Finegan B, Martínez C,
Mendonza C, Villalobos R, Medellín C, Villanueva C, Mendoza T, Aguilar A, Padilla D
(2015) Climate-Smart Territories (CST): an integrated approach to food security, ecosystem
services, and climate change in rural areas. In: Minang PA, van Noordwijk M, Freeman OE,
Mbow C, de Leeuw J, Catacutan D (eds) Climate-smart landscapes: multi-functionality in
practice. World Agroforestry Center (ICRAF), Nairobi

Myuri B, Obert J, Mafongoya PL (2017) Integration of indigenous and scientific knowledge in
climate adaptation in Kwazulu-Natal, South Africa. Change Adaptation Socioecol Syst De
Gruyter 3:56–67

National Bureau of Statistics-NBS (2016) Nigeria gross domestics product report. Federal Government of Nigeria

Okali C (2012) Gender analysis: engaging with rural development and agricultural policy processes.
Working   paper.   https://assets.publishing.service.gov.uk/media/57a08a7fe5274a31e000062a/
FAC_Working_Paper_026.pdf PRESANCA (2011) Centroamérica en Cifras. Datos de seguridad
alimentaria nutricional y agricultura. http://www.fao.org/fileadmin/user_upload/AGRO_Noticias/
docs/CentroAm%C3%A9ricaEnCifras.pdf

Raymond MC, Fazey I, Reed MS, Stringer LC, Robinson GM, Evely AC (2010) Integrating local
and scientific knowledge for environmental management. J Environ 91:1766–1777

United Nations Development Program-UNDP (2014) Human development report 2014 Sustaining
human progress: reducing vulnerabilities and building resilience. UNDP, New York

United States National Aeronautics and Space Administration -NASA (2016) Climate trends
continue to break records. July 19, 2016 – http://www.nasa.gov/feature/goddard/2016/climate-
trends-continue-to-break-records. 1 Oct 2016

Williams PA, Crespo O, Abu M (2019) Adapting to changing climate through improving adaptive
capacity at the local level-the case of smallholder horticultural producers in Ghana. Clim Risk
Manag Elsevier 23:124–135

10

Climate Change Adaptation Strategies Among Cereal Farmers in Kwara State, Nigeria

S. A. Aderinoye-Abdulwahab and T. A. Abdulbaki

Contents

Abstract

Agriculture is the art and science of food production which spans soil cultivation, crop growing, and livestock rearing. Over the years, it has served as a means of employment and accounts for more than one-third of total gross domestic product. Cereals, which include rice, maize, and sorghum, are the major dietary energy suppliers and they provide significant amounts of protein, minerals (potassium and calcium), and vitamins (vitamin A and C). The growth and good yield of cereal crop can be greatly influenced by elements of weather and climate such as temperature, sunlight, and relative humidity. While climate determines the choice of what plant to cultivate and how to cultivate, it has been undoubtedly identified as one of the fundamental factors that determine both crop cultivation and

S. A. Aderinoye-Abdulwahab (✉) · T. A. Abdulbaki
Department of Agricultural Extension and Rural Development, Faculty of Agriculture, University of Ilorin, Ilorin, Nigeria
e-mail: aderinoye.as@unilorin.edu.ng

livestock keeping. The chapter, though theoretical, adopted Kwara State, Nigeria, as the focus due to favorable weather conditions that support grains production. It was observed that the effect of climate change on cereal production includes: drastic reduction in grains production, reduction in farmers' profit level, increment in cost during production, diversification to nonfarming activities, and discouragement of youth from participating in agricultural activities. Also, the adopted coping strategies employed by farmers in the focus site were early planting, planting of improved variety, irrigation activities, alternates crop rotation, and cultivation of more agricultural areas. The chapter thus concluded that climate change has negative impact on cereals production and recommends that government should provide communal irrigation facilities that will cushion the effect of low rains on farmers' productivity, while early planting and cultivation of drought-resistant cultivars should be encouraged.

Keywords

Grains · Crop rotation · Early planting · Irrigation · Yield

Agriculture and Climate Change

Our ancestors advanced from being hunter gatherers to live a more settled lifestyle since about a thousand years ago, residing in one place for a lengthy period rather than living a nomadic lifestyle. This change happened concurrently with the taming of different species of plants and animals, which were bred, kept or planted in close proximity rather than hunted. Over the periods, agriculture continued to advance although, up until the twentieth century, most of the work were undertaken by mankind and animals rather than machines. Such fauna and flora activities (which could also be natural), including human's method of cultivation as well changes in weather and climate, left the environment in a tough situation that we now refer to as climate change and its impact. Agriculture is largely reliant on weather and climate, hence climatic factors such as rainfall, sunlight, airstream, and temperature influence the distribution and productivity of crops (Sokoto et al. 2016).

Of all human endeavors, agriculture remains one of the most demanding on land. Land is thus in constant short supply given the effects of biophysical factors such as rain, temperature, and topography of an area, as well as unsustainable land management practices which include deforestation, uncontrolled soil nutrient mining, and cultivation on steep slopes (FAO 2015). The implication of such effects on land may comprise overgrazing, desertification, soil erosion, deforestation, among others and all of these further translate into shortage of land for human's endeavors. Perhaps the continued presence of these effects is one of the reasons why scholars have maintained that the ozone layer may eventually be depleted given that the earth is warming up (IPCC 2007; Mboera et al. 2012; FAO 2015) as a result of people's activities which enable greenhouse gases (GHG) like methane, water vapor, nitrous oxide, and carbon dioxide to trap heat into space.

Climate, a manifestation of weather and other atmospheric conditions, has largely been acknowledged as one of the fundamental indicators that determine both crop cultivation and livestock keeping. It is a long-term average weather conditions that directly or indirectly impact the events that play out on farms. Climatic conditions of any environment govern the choice of crop and how they will be planted, as well as the yields and nature of livestock to keep in a given location. This statement is alluded to in the work of Ajadi et al. (2011) where it was similarly shown that biophysical factors such as sunlight energy, temperature, humidity, and other climatic variables determine the global distribution and yield of crops and livestock. On the other hand, climate change is the long-term change in average temperature as a result of the earth's warming which may eventually translate into the depletion of the ozone layer. Although, an inconsistent phenomenon, climate change points clearly at threats to the future of people through the depletion of the ozone layer, thereby negatively impacting on all segments of human endeavor, agriculture inclusive. As a result of human and natural activities that release greenhouse gases (GHG) which had hitherto trapped heat within the atmosphere (Mboera et al. 2012), higher concentrations of these gases in the atmosphere leads to the earth's warming. These human activities could include but not limited to: use of fossil fuel, land use change, and agriculture (IPCC 2007; Mboera et al. 2012; Omumbo et al. 2011). Many people in developed regions prefer to assume that climate change is not real than to prepare to adjust their living standards to suit the effect of climate change (Mboera et al. 2012). Hence, there is a need to increase farmers' awareness on the effect of climate change as well as adequately prepare for its mitigation. Researches to proffer ease of living with climate change and its impact are therefore expedient.

It has equally been shown in another instance that the planet's temperature is increasing, leading to frequent changes in rainfall patterns and extreme events such as droughts (Zoellick 2009). The International Panel of Climate Change (IPCC 2007) also submitted that agriculture, forestry, and the change in land use, account for up to 25% of human induced GHG emissions while agriculture is the chief source of released methane and nitrous oxide, bearing in mind that these two gases constitute the substantial part of GHGs. Similarly, Odjugo (2010) stated that climate change is undeniable and its impacts are clearly seen in rising temperatures, low rainfalls, desertification, water scarcity, health, and agricultural problems. It has also been reported that several million hectares of crops and livestock have been destroyed due to climate-induced tragedies (FAO 2015). Climate change has therefore impacted on the economy of nations so much so that agriculture sector absorbs about 22% of the entire damage in developing countries (FAO 2015). Therefore, there is a dire need for farmers to be aware of the extent of destruction that the impact of climate change can constitute in their livelihoods as this will work to boost their preparedness for its mitigation. Reuben and Barau (2012) had similarly reported that incidences of low humidity during propagative stage will definitely result in poor yields of crops while the Food and Agriculture Organization (FAO) also poised that the crop sector is the most affected during unfavorable weather episodes (FAO 2015).

It is imperative to note that countries would feel the impact of climate change differently; while the effect may be positive in some areas, other climes might be negatively affected. Some additional degree of climate change and its impact is unavoidable regardless of the amount of measures put in place to ameliorate or forestall it. As such, there is an inevitable need to adapt farming to resilient conditions of agricultural systems. The two major means of reducing climate impact are mitigation and adaptation. Adapting agriculture to climate change and sustaining adequate food production could help to resolve the current glitches. However, with growing levels of GHGs, climate change and its consequences will continue to pose new challenges as well as mete out untold hardship on people. Currently, agriculture (and related sectors) contributes about a quarter of human-induced GHG emissions (IPCC 2007; FAO 2015), so it is highly necessary to reduce these emissions and other unfavorable effects from our environment. Many times, decreasing resource inputs and increasing efficacy goes hand in hand with mitigating emissions. If we can reduce the concentration of GHGs in our atmosphere, then extreme climate circumstances in the future would have been reduced considerably, and thus, it will be easier to adapt to climate change.

This chapter will therefore dwell on the interconnectedness of agriculture, climate change, and the effects of climate change on grain production. It will as well highlight grain farmers' coping and adaptation strategies to the impact of climate change.

Focus Study Site

Kwara State is located in a hot and humid region with two major seasons; it has tropical wet and dry climate (Olanrewaju 2009), each lasting for about 6 months. Kwara State has an annual rainfall range of 1000–1500 mm from March until early September, although this is gradually shifting forward due to climate variations; while the dry season used to be from October to March. Again, this is also on the moving trend as temperature continues to be on the high side up until April/May. Temperature is uniformly high and ranges between 25 °C and 30 °C in the wet season; while the dry season ranges between 33 °C and 34 °C. On average, the state is occupied by subsistence tuber, legume, and cereal farmers who depend solely on rain-fed agriculture; hence one can only imagine how climate change would impact on these farmers. The major occupation of the people of Kwara is largely agricultural-related as more than 70% of the population are farmers. Moreover, the plain to slightly gentle topography of Kwara State further makes it favorable for agricultural growth while other climatic patterns, vegetation, and the fertile soil make the state suitable for the farming of an extensive range of food crops such as cereals, cowpea, cassava, and tree crops, such as cashew and mango (Olanrewaju 2012). This review adopted Kwara State as a focus site because there are large number of cereal farmers in addition to favorable soil condition for production of grains such as rice, maize, millet, and sorghum, as this can favor low consumption of nutrient, low managerial procedure, as well as two planting seasons for rice and maize among others.

Agriculture and Cereal Production in Nigeria

Nigeria is an agrarian country with a reasonable percentage of its population engaging in agriculture at subsistence level (Adhvaryu et al. 2019). Like most countries in Sub-Saharan Africa (SSA), farmers in Nigeria not only live in rural areas, but their agricultural holdings are generally minute with about two-thirds of the total production being practiced with the use of simple methods and the bush-fallow system of cultivation. Agriculture, as practiced widely in Nigeria and other developing countries, is highly subjected to climate variability, and Nigeria's comprehensive range of climate differences supports a broad variety of the production of food crops (Olanrewaju 2012). Of the vast area of land for cultivation in Nigeria, cereals like sorghum, millet, rice, and corn are predominantly grown, but maize has overtaken the other traditional cereals such as wheat in terms of cultivation and popularity. Grains are rich in starches and contain substantial amounts of protein, as well as some fat and vitamins. They are essentially the world's staple food with over 70% of the world's harvested area producing a billion and a half tons of grains annually (Sokoto et al. 2016). Despite this, cereal production is set for a global increase with wheat, maize, and barley accounting for most of the rise in cereal production, while rice production is likely to maintain the increase as produced in 2018 (FAO 2019).

Rice is a starchy cereal that contains fiber, protein, vitamins, and minerals. It is consumed by more than half of the world's population (Rathna Priya et al. 2019) and is increasingly an important crop in Nigeria, with some species growing within 3 months. Africa cultivates about half of the 14 million metric tonnes of the annual global rice production (FAO 2019), while Nigeria is the largest producer and the leader in terms of consumption as well as importation in Africa. Nigeria's production has steadily advanced from two million metric tonnes to 8.0 million metric tonnes between 2008 through 2018 with the aim of reaching 18 million tonnes by 2023 (FAO 2019). This is due to rice policy reforms by the Federal Government of Nigeria who in its bid to eliminate rice importation enacted the anchor borrower's scheme that have paved way for proliferation of rice farming in the country (FAO 2019). Rice is not usually grown in isolation but with other crops such as sorghum, maize, and sweet potato. Initially, rice was noted as a luxurious food but has fast become a regular diet in every home in the country.

In Nigeria, maize is equally a significant food and feed crop which maintains an essential status for rural food security (Stevens and Madani 2016). Introduced to Nigeria in the sixteenth century, maize is the fourth most utilized cereal ranked below sorghum, millet, and rice (FAOSTAT 2014). Maize production in Nigeria is above ten million metric tons (Mundi Index 2018; FAO 2019), and it accounts for almost half of the calories and protein consumed in Eastern and Southern Africa (ESA), and one-fifth of the calories and protein consumed in West Africa. It is a significant source of carbohydrate, protein, iron, vitamin B, and minerals. Additionally, maize is essential in the cropping system of the small-scale Nigerian farmer and is frequently grown in mixture with other crops such as legumes or even cereals under traditional practice. This is not only because maize is relatively more afford-able than other cereal crops, but it is also widely accepted in the region as edible plant (Sowunmi and Akintola 2010). Apart from millet, maize is the earliest grain that is harvested in any given season in Nigeria, hence offering respite to the usual food scarcity being experienced at such periods of the year while maize consumption is generally acceptable in various forms across the country (Mundi Index 2018).

It is now a common knowledge that farming activities, including cereal produc-tion, has put global populations at vulnerable climatic situations and many studies have evolved in the area (Adger et al. 2004). In Nigeria, the vulnerability to climate change can be seen in the overwhelming effects of recent climatic catastrophes in all the six geopolitical zones of the nation. The protracted droughts resulting in increased temperatures as currently being observed in the northern region and the late arrival and early cessation of rain (Apata et al. 2009) are manifestations of climate change impact. It is evidently clear that Nigeria is highly susceptible to the effect of climate change because of its wide (800 km) coastline which is prone to sea level rise and the various challenges related to the violent storms (Apata et al. 2009). In view of the foregoing, it has become expedient to assess the various adaptation strategies adopted by cereal farmers in Nigeria in order to better cope with the effect of climate change as well as serve as knowledge bridging gap for other farmers to leverage on. The next sections would highlight the effect of climate change on cereal production before drawing conclusion and recommendations that cereal farmers in the region where Kwara State is located can derive ideas from.

The Impact of Climate Variability on Agriculture in Nigeria

Climate variability may have an indirect influence on the goal of poverty allevi-ation in developing countries (Skoufias et al. 2011) as efforts at reducing the variations in climate help to increase agricultural yields which ultimately translate into better incomes for farmers and improved food security for the populace. This suggests that there might be efforts from development experts that could nurture linkages between climate vulnerabilities and development policies. Perhaps, such efforts and policy debates could eventually address the impact of climate change on agricultural-related practices. Falling incidences of climate variability will cushion the vulnerability levels of farmers (Skoufias et al. 2011), and the effect of climate change will normally be felt either directly or indirectly on all segments of the environment to include: socio-economy, water resources, food security, human health, ecosystems, coastal zones, and other related sectors. Fluctuations in precipitation and melting of glaciers are pointers to acute water shortages, soil erosion, and/or flooding (IPCC 2007), and these will absolutely affect agriculture produce. These climatic variations would inadvertently result in crop losses and low profits for farmers.

The steady increase in temperatures has necessitated an unavoidable shift in crop growing seasons for farmers in Nigeria and this has affected sufficient food availability and variations in the spread of disease vectors (Odjugo 2010;

Apata et al. 2009). The variations in climate amplifies vulnerability of people to diseases through the disease-carrying vectors. Moreover, upsurge in temperatures can possibly result in increased rates of extinction of many species and habitats. This, in itself, is not favorable for the ecosystem, and in a like manner, pest and diseases migrate in response to climate change with serious consequences on agriculture and the entire planet (IPCC 2007; Earth Journalism Network 2016). Such irregular temperature patterns and movements of pests lead to erratic rainfall and sunshine patterns, thereby resulting in crops' instability. It is therefore apparent that there is a heightened (above global mean) warming rate on the lands as evaporative cooling has greatly reduced, and there exists a smaller thermal inertia when compared to those found in the oceans. Boko et al. (2007) reported a similar scenario for Nigeria where the authors stated that the warming rate in Nigeria was above the mean rate for Africa.

Extreme rainfall patterns and variations in Nigeria have become a serious production risk for agricultural production system that is largely rain-fed (Olayide et al. 2016). The rain-fed agricultural production system is vulnerable to seasonal variability which affects the livelihood outcomes of farmers who greatly depend on this system of production. Previous studies have also shown that varying temperature fluctuations, heat waves, relative humidity, and rainfall are bound to compromise agricultural production as substantial losses are observed in the yield of grains such as rice, millet, sorghum, and maize in Nigeria (Sokoto et al. 2016). It was similarly found that maize along with other cereals were more susceptible to climate variability with untold consequence on the yield (Bismark and Richard 2019). Similarly, it was found that yields and productivity of maize, millet, sorghum, and rice were negatively affected by unfavorable climatic conditions such as drought, excessive temperature, and low rainfall (Bamiro et al. 2020).

The complications that are linked to climate change are not the same across the country. Nigeria, for example, has a tropical climate with two precipitation regimes spanning the north and south of the equator. The north is typically low in precipitation and high in temperature while parts of southwest and southeast are distinguishingly lower in temperatures with high amount of rainfalls (Nkechi et al. 2016; Akande et al. 2017). Obviously, the northern climate can lead to grave ecological effects as demonstrated by aridity, drought, and desert encroachment in the north, and flooding and erosion in the South (Nkechi et al. 2016; Akande et al. 2017). It was further shown through vulnerability analysis that States in the north, especially north east and north west region, experience higher degrees of vulnerability to climate change than those in the south (Federal Ministry of Environment 2014; Madu 2016). These variations in the northern climate are further compounded by desertification, loss of the wetlands, with a disturbing decrease in the volume of surface water, flora, and fauna resources on land (Abdulkadir et al. 2017; Ebele and Emodi 2016). Interestingly, the Southwest and Southeast are relatively less vulnerable when compared to other parts of the country because of the eco-friendlier climate patterns as depicted by high precipitation and relative humidity. Although, despite the relatively less vulnerability of the southern region of the country, the South-South (Niger Delta region) is most susceptible, due to sea level rise, increased rainfall, coastal destruction, and flooding, which has resulted in the dislodgment of many settlements (Matemilola 2019).

The form of vulnerability to climate change usually corresponds with agronomic activities in a given location (Madu 2016). The argument here is that the northern regions of Nigeria, which coincidentally have higher degrees of rurality, are more vulnerable to climate change (Madu 2016). Hence, the adoption of existing and new technologies for adapting to climate change and variability is a high priority for most part of Nigeria. Evidence from secondary data has shown that farmers in the focus study site (Kwara State is located in the northern part of Nigeria) are already feeling the impact of climate change in many ways (Sokoto et al. 2016). The effect of climate change on cereals production is firstly noticeable in a sharp increase in production costs and reduced grains production while this translates into decreases in farmers' profit level (Sokoto et al. 2016). Ordinarily, the cost during production could have been minimal and relatively stable but for the heightened climate variabilities. In developing countries including Nigeria, climate variability and change do encourage heat and moisture stresses thereby adding to an already long list of existing problems (Earth Journalism Network 2016). How then are countries in the tropical regions able to deal with climate change impacts?

Climate Change Adaptation Strategies for Cereal Farmers in Nigeria

Adaptation is a method through which people make themselves better able to cope with an uncertain future. Adapting to climate change will therefore entail the application of efficient actions to lessen the adverse effects of climate change (or exploit the positive ones) by making the necessary modifications in order not to greatly feel the negative impact. Emerging countries, such as Nigeria, typically perceives industrialized countries as climes with reduced vulnerability and better adaptation strategies given that such regions are better able to realize the prospects in cold weather episodes and hence make calculated moves to strengthen their agricultural production (Achike and Onoja 2014). The international community through the UNFCCC are in serious deliberations to find an effective means to battle climate change. It therefore becomes pertinent for the international community to entrench, in their future decisions, processes that will assist developing nations with transfer of knowledge, technology, and financial resources to adapt at all levels and in all sectors.

Recent study offered that farmers in Africa use crop diversification to build resilience in the agriculture sector (Mango et al. 2018), although this method may not be a favorable means among certain other farmers. However, adaptation, regardless of whatever form it takes, is already considered a major and important integral part of any future climate change regime. Some of the known technologies in use for climate change variability and impacts include: crop diversification, the adoption of drought-tolerant and early-maturing varieties of crops, and planting of cover crops (Federal Ministry of Environment 2014; Achike and Onoja 2014). Studies have shown that farmers in Nigeria are being supported by government and other non-governmental organizations to better adapt to climate change using these, as well as other methods to deal with climate change impact (Ifeanyi-obi and Nnadi 2014). In addition, relevant weather-related information and skills training that can enhance productivity can be offered by agricultural extension services (Akintonde and Shuaib 2016). Although the current irregularity of extension services in Nigeria is a limitation to the adaptation strategies (Akintonde and Shuaib 2016).

Other adaptation strategies used by farmers in SSA including Nigeria are: early planting of crops, a condition where crops would have enjoyed some reasonable amount of rain before it ceases. Rainy season, of the current year, in certain parts of Nigeria experienced a somewhat climate change, and this means that Kwara State, along with many other states in the country, did not enjoy rains in the months of July and August; supposedly, those months should have been the peak of rains when activities for farmers would normally climax. Thus, some farmers who tapped into the opportunity of early planting sowed particular grains in May/June (maize for instance) when there were initial heavy rains and those farmers were able to survive the drought occasioned by the lack of rains. However, most grains that would normally be planted later in July did not survive as this coincided with the cessation of rains in July and August and farmers lost their resources (monies, inputs, time, and energy).

Improved variety is another adaptation strategy in use in order to cope with climate change impact. Farmers now plant improved varieties that have high resistance to pest and diseases, requires low water, matures quickly, and with many more qualities. This strategy was found to be popular among cereal farmers in Nigeria. For example, close observation of rice farmers in Kwara State showed that they employed similar means to combat the impact of climate change within the region. Farmers now plant a locally crossbred rice variety from Kebbi State that germinates in 12 weeks while some other farmers are yet to tap into this innovation and are still growing older rice varieties that reach maturity in 4 months and above. The improved rice variety offers farmers the opportunity to cultivate and harvest rice twice in one growing season especially if such farmer leveraged on early planting strategy. The snag here, however, is that the improved seeds are more expensive, although on the flipside, farmers can produce seeds for the next growing season from the current one and can consequently save some costs in the next planting season. Again, this method may not work if prolonged rain shortages are experienced and there are no facilities for irrigation.

Crop rotation is also one of the strategies used by the farmers to assuage the impact of climate change. Farmers in SSA including Nigeria do not restrict themselves to sowing a mono-crop. They use grains and/or legumes to rotate across seasons. A farmer might plant corn along with other tuber crops such as cassava in one season, and then proceed to corn and cowpea in another season. There are times when farmers are able to predict what the next planting season would look like given their experiences from previous climate variation patterns and based on local climatic indicators, although their predictions may not always work as projected. For instance, farmers may envisage a favorable climate in the next growing season if temperatures in the current year are on the high and rains were too scanty. In such a situation, a farmer can predetermine which crops would offer more yield with high precipitation and less vulnerability to such weather. At the return of the rains in Kwara State at the tail end of August and beginning of September this year, farmers who have suitable lands hastened to plant soybean as this crop require less amount of rainfall and would normally germinate within 12 weeks when it is expected that the little rain that would fall in the remaining part of the season would be enough to see the crop to harvest in November. At some other times, farmers may rely on weather forecasts from relevant government agencies to determine what the forthcoming growing season would look like and adequately prepare for the crops that would thrive better in such weathers. Such information can also be leveraged on to determine what crops offer relative advantage over others, in order to make decisions as to the crops to rotate with the other in subsequent seasons. Such calculations help the farmers to plan better as well as make calculated risks that can translate to reduced climate change impacts.

In addition, farmers in Kwara State have also applied crop diversification as another adaptation strategy. For example, farmers have changed the crops grown on previously cultivated lands from a particular cereal crop to another due to climatic changes. Land that was previously being farmed for rice production because of abundance of water have now been converted to cultivation of maize and sorghum after such lands have lost their water content due to enduring droughts. The adaptation strategies adopted by cereal farmers in Kwara State are as seen in similar works where it was reported that crop diversification, adoption of drought-tolerant cultivars, early-maturing varieties of crops, and planting of cover crops are among the adaptation strategies being employed by farmers in Nigeria (Federal Ministry of Environment 2014; Achike and Onoja 2014).

Irrigation and drainage activities were also adaptive measures that farmers in Nigeria usually put into consideration during planting operations in order to secure water for plants during dearth periods (for example, as seen in northern Nigeria) and also mitigate the impact of flooding should the need arise, especially in the coastal southern region of Nigeria. This will reduce crop loss during insufficient and excessive rainfall like the full 2 months break earlier described in the current rainy season in Nigeria. Secondary data collected, observations and experiences from previous and current planting seasons showed that farmers in northern Nigeria had relied on strategies like direct interventions such as dike building to prevent flooding, large-scale relocation of farmers from excessively eroded areas, new crop selection, building of dams to expand irrigation, and crop rotation in order to curb the menace of climate change. Methods such as this would also ensure adequate yield of crops by farmers. Rice farmers in Kwara State, for instance, constructed make-shift dams at several points on their farms, where rain water is usually collected/deposited for use during dearth periods. Unfortunately, such water gets dried up whenever there is protracted or intense water shortage and farmers will sometimes need to walk long miles to areas where there are other sources of water (wells and boreholes) in order to irrigate their farms. The current prolonged lack of rains in Kwara State led to huge losses as farmers were not hitherto prepared and consequently could not cope well with the situation.

Another adaptation strategy adopted by cereal farmers in the focal site is cultivation of more agricultural areas. By this, it was observed that farmers chose not to wait until harvest period before they will discover heavy crop losses and/or damages; hence they became proactive by cultivating large expanse of lands to increase their productivity. Some farmers have increased their farm sizes through cultivation of more lands as this will make them generate more income as well as improve their livelihood. Interestingly, majority of farmers in Kwara State, like other parts of the country, do not own the lands where they farm. First of all, all lands in the country belong to the government and anyone who intends to take possession would have to seek the owner's (government) permission in the form of "registration of title" which attracts special charges. However, farmers have usually cultivated lands without the owner's permission and they are always ready to move to other expanse of lands whenever the rightful owners emerge to take over the lands. Farmers have occasionally recorded huge losses when such circumstances arise but they would rather take such risks as it has almost become a norm among them. It is only the large-scale commercial farmers that normally acquire the lands rightfully. Meanwhile, while farmers may be increasing their yields by cultivating more expanse of lands, this method is indirectly expanding land used for agricultural activities and will consequently increase their risks and exposure to climate change impacts.

Livelihood diversification is also one strategy being used by cereal farmers in the focus site. Some farmers who feel they could no longer cope with the climate change impacts have decided to change the means/source of livelihood. Some of the cereal farmers in Kwara State have attempted diversification to nonfarming activities as adaptation strategies while discouraging youths, who are becoming more interested in white collar jobs, from participating in agricultural activities. Hence, though sadly, some of the farmers have boycotted farming for artisan jobs as well as bicycle riding jobs in order to sustain their families.

Conclusion

Agricultural activities play a major role in the development of any country, and the effect of climate change on agronomic activities cannot be underrated due to its direct impact on agricultural production. Climate change may be a global issue but its effects will automatically differ between geographical regions. While food production in some areas may suffer from extreme weather and temperatures, farming in other regions might benefit from longer growing seasons and warmer climates. Early planting, crop diversification, planting of improved variety, irrigation and drainage practices, crop rotation activities, and cultivation of more agricultural land areas were the adopted strategies used in the focus site to help farmers cushion the impact of climate change. Therefore, farmers need to continue to adopt those strategies to combat this situation, in order to keep sustaining their livelihoods.

Recommendations

Based on observations and inferences highlighted, cereal farming, in Nigeria as a whole and particularly in the area of focus, would be enhanced if these recommendations are taken:

1. There is a need for meteorological agencies (for example, Nigerian meteorological agency – NiMet) to widen their scope and expand their methodology to ensure that small-scale farmers benefit from their forecasts, in order to reduce farm losses due to unfavorable weather episodes. This will make farmers to predetermine the crops and time to plant.
2. Rain-water collection systems can be made available by collaborative efforts of the farmers, while government and other stakeholders serve as buffer during dry season and early cessation of rainfall. This will help in situations like the one experienced this current season by farmers in Kwara State where farmers became helpless as a result of unexpected shortage of rains.
3. The public sector (government) can further encourage small-scale farmers by instituting favorable policies that will stimulate farmers to remain on the fields as well as enhance them to better cope with unfavorable climatic conditions. Such policies, like fertilizer subsidy and those enacted for increased rice production, would reduce cases of livelihood diversification where youths are seen to migrate to cities for alternative jobs.
4. Provision of communal irrigation facilities (every problem must not be solved at the instance of government) for use in times of water scarcity to avoid food shortages for the populace. This will also reassure farmers that their efforts would not be a waste and the fear of unexpected huge losses on their part can be allayed.

References

Abdulkadir A, Lawal MA, Muhammad TI (2017) Climate change and its implications on human existence in Nigeria: a review. Bayero J Pure Appl Sci 10(2):152–158. https://www.ajol.info/index.php/bajopas/article/viewFile/170772/160195

Achike AI, Onoja AO (2014) Greenhouse gas emission determinants in Nigeria: implications for trade, climate change mitigation and adaptation policies. https://www.journalijecc.com/index.php/IJECC/article/view/27273/51196

Adger WN, Brooks N, Bentham G, Agnew M, Eriksen S (2004) New indicators of vulnerability and adaptive capacity. Tyndall centre technical report, no 7. Tyndall Centre for Climate Change Research, University of East Anglia, Norwich

Adhvaryu A, Nyshadham A, Tamayo J (2019) Managerial quality and productivity dynamics. Harvard Business School. Working paper 19–100

Ajadi BS, Adeniyi A, Afolabi MT (2011) Impact of climate on urban agriculture: case study of Ilorin City, Nigeria. Global J Human Soc Sci 11(1)., Type: Double Blind Peer Reviewed International Research Journal Publisher: Global Journals Inc. (USA)

Akande A, Costa AC, Mateu J, Henriques R (2017) Geospatial analysis of extreme weather events in Nigeria (1985–2015) using self-organizing maps. Adv Meteorol. https://doi.org/10.1155/2017/8576150

Akintonde JO, Shuaib L (2016) Assessment of level of use of climate change adaptation strategies among arable crop farmers in Oyo and Ekiti States, Nigeria. J Earth Sci Clim Chang 7:369.

https://www.omicsonline.org/open-access/assessment-of-evel-of-se-of-climate-change-adapta
tion-strategiesamong-arable-crop-farmers-in-oyo-and-ekiti-states-nigeria-2157-7617-1000369.
php?aid=80244

Apata TG, Samuel KD, Adeola AO (2009) "Analysis of climate change perception and adaptation
among arable food crop, farmer in South Western Nigeria" Contributed paper at the International Association of Agricultural Economists 2009 conference, Beijing

Bamiro O, Adeyonu A, Ajiboye B, Solaja S, Sanni SO, Faronbi O, Awe T (2020) Effects of climate
change on grain productivity in Nigeria (1970–2014). IOP conference series. Earth Environ Sci
445:012058. https://doi.org/10.1088/1755-1315/445/1/012058

Bismark MB, Richard K (2019) Climate Variability Effect on Food Crop Yield among the
Smallholder Farmers in Lower Off in River Basin, Ghana. Journal of Agriculture and Environmental Sciences 8(2):66–74. Published by American Research Institute for Policy Development. https://doi.org/10.15640/jaes.v8n2a9

Boko M, Niang I, Nyong A, Vogel C, Githeko A, Medany M, Osman-Elasha B, Tabo R, Yanda P
(2007) In: Parry ML, Canziani OF, Palutikof JP, van der Linden PJ, Hanson CE (eds) Africa
climate change: impacts, adaptation and vulnerability. Contribution of working group II to the
fourth assessment report of the intergovernmental panel on climate change. Cambridge University Press, Cambridge, UK, pp 433–467

Earth Journalism Network (2016) Introduction to climate change. Available at: https://
earthjournalism.net/resources/introduction-to-climate-change

Ebele NE, Emodi NV (2016) Climate change and its impact in Nigerian economy. J Sci Res Rep
10(6):1–13. http://www.journaljsrr.com/index.php/JSRR/article/view/21917/40737

FAO (2015) Climate change and food systems: global assessments for food security and trade. In:
Elbehri A (ed) A publication of the Food and Agricultural Organization, Rome. Avaliable at:
http://www.fao.org/3/a-i4332e.pdf

FAO (2019) Food outlook – biannual report on global food markets. Rome. Licence: CC BY-
NC-SA 3.0 IGO

FAOSTAT (2014) The state of food and Agriculture, investing in Agriculture for a better future.
http://Faostat.fao.org

Federal Ministry of Environment (2014) United nations climate change Nigeria. National Communication (NC). NC 2. 2014. https://unfccc.int/sites/default/files/resource/nganc2.pdf

Ifeanyi-obi CC, Nnadi FN (2014) Climate change adaptation measures used by farmers in south-
South Nigeria. J Environ Sci Toxicol Food Technol 8(4). https://pdfs.semanticscholar.org/a63d/
22bf8b8ebde892a7c9d761ee1653e0e11df6.pdf

IPCC (2007) Climate change 2007: impacts adaptation and vulnerability. In: Contributions of
working group II to the fourth assessment report of the intergovernmental panel on climate
change. Cambridge University Press, Cambridge, UK

Madu IA (2016) Rurality and climate change vulnerability in Nigeria: assessment towards evidence
based even rural development policy. Paper presented at the 2016 Berlin conference on global
environmental change, 23–24 May 2016 at Freie Universität Berlin. https://pdfs.
semanticscholar.org/508b/94cab07b84a703b44eca1089326cc98d7495.pdf?_ga=2.154518008.
112403230.1572433568-162569160.1557482164

Mango N, Makate C, Mapemba L et al (2018) The role of crop diversification in improving
household food security in central Malawi. Agric Food Sec 7:7. https://doi.org/10.1186/
s40066-018-0160-x

Matemilola S (2019) Mainstreaming climate change into the EIA processing Nigeria: perspectives
from projects in the Niger Delta region. Climate 7(2):29. https://doi.org/10.3390/cli7020029

Mboera L, Mayala B, Kweka E, Mazigo H (2012) Impact of climate change on human health and
health systems in Tanzania: a review. Tanzan J Health Res 13. https://doi.org/10.4314/thrb.
v13i5.10

Mundi Index (2018) Nigeria country profile. Available at: https://www.indexmundi.com/nigeria/
agriculture_products.html

Nkechi G, Onah N, Ali A, Eze E (2016) Mitigating climate change in Nigeria: African traditional
religious values in focus. Mediterr J Soc Sci 7(6):299–308. https://www.mcser.org/journal/
index.php/mjss/article/view/9612

Odjugo PAO (2010) Adaptation to climate change in the agricultural sector in the semi-arid region of Nigeria. Paper presented at the 2nd international conference: climate, sustainability and development in semi-arid regions, Fortaleza-Ceará, Brazil, August 16–20, 2010

Olanrewaju RM (2009) Climate and the growth cycle of yam Plant in the Guinea Savannah Ecological Zone of Kwara state, Nigeria. J Meteorol Clim Sci:43–48

Olanrewaju RM (2012) Effect of climate on yam production in Kwara State, Nigeria. Environmental Issues, Department of Geography and Environmental Management, University of Ilorin, Ilorin 3(1):30–34

Olayide OE, Kow Tetteh I, Popoola L (2016) Differential impacts of rainfall and irrigation on agricultural production in Nigeria: any lessons for climate-smart agriculture? Agric Water Manag 178:30–36. https://doi.org/10.1016/j.agwat.2016.08.034. ISSN 0378-3774. Available: http://www.sciencedirect.com/science/article/pii/S0378377416303286

Omumbo JA, Lyon B, Waweru SM, Connor SJ, Thomson MC (2011) Raised temperatures over Kericho tea estates: revisiting the climate in the east African highlands malaria debate. Malar J 10:2

Rathna Priya T, Eliazer Nelson ARL, Ravichandran K (2019) Nutritional and functional properties of coloured rice varieties of South India: a review. J Ethn Food 6:11. https://doi.org/10.1186/s42779-019-0017-3

Reuben J, Barau AD (2012) Resource use efficiency in yam production in Taraba state, Nigeria, journal of agricultural science. Kamla-Raj 3(2):71–77

Skoufias E, Vinha K, Conroy HV (2011) The impacts of climate variability on welfare in rural Mexico, policy research working paper 5555. World Bank, Washington, DC

Sokoto M, Tanko L, Abubakar L, Dikko A, Abdullahi Y (2016) Effect of climate variables on major cereal crops production in Sokoto state, Nigeria. Am J Exp Agric 10:1–7. https://doi.org/10.9734/AJEA/2016/20020

Sowunmi FA, Akintola JO (2010) Effect of climatic variability on maize production in Nigeria. Res J Environ Earth Sci 1:19–30

Stevens T, Madani K (2016) Future climate impacts on maize farming and food security in Malawi. Sci Rep 6:36241

Zoellick RB (2009) A climate smart future, the nation newspapers. Vintage Press Limited, Lagos, p 18

11

Impact of Moisture Flux and Vertical Wind Shear on Forecasting Extreme Rainfall Events in Nigeria

Olumide A. Olaniyan, Vincent O. Ajayi, Kamoru A. Lawal and Ugbah Paul Akeh

Contents

O. A. Olaniyan (✉) · U. P. Akeh
National Weather Forecasting and Climate Research Centre, Nigerian Meteorological Agency,
Abuja, Nigeria

V. O. Ajayi
West African Science Service Center on Climate Change and Adapted Land Use, Federal
University of Technology, Akure, Ondo State, Nigeria

Department of Meteorology and Climate Science, Federal University of Technology, Akure, Nigeria
e-mail: voajayi@futa.edu.ng

K. A. Lawal
National Weather Forecasting and Climate Research Centre, Nigerian Meteorological Agency,
Abuja, Nigeria

African Climate and Development Initiative, University of Cape Town, Cape Town, South Africa

Abstract

This chapter investigates extreme rainfall events that caused flood during summer months of June–September 2010–2014. The aim is to determine the impact of horizontal moisture flux divergence (HMFD) and vertical wind shear on forecasting extreme rainfall events over Nigeria. Wind divergence and convective available potential energy (CAPE) were also examined to ascertain their threshold values during the events. The data used include rainfall observation from 40 synoptic stations across Nigeria, reanalyzed datasets from ECMWF at $0.125° \times 0.125°$ resolution and the Tropical Rainfall Measuring Mission (TRMM) dataset at resolution of $0.25° \times 0.25°$. The ECMWF datasets for the selected days were employed to derive the moisture flux divergence, wind shear, and wind convergence. The derived meteorological parameters and the CAPE were spatially analyzed and superimposed on the precipitation obtained from the satellite data. The mean moisture flux and CAPE for some northern Nigerian stations were also plotted for 3 days prior to and 3 days after the storm. The result showed that HMFD and CAPE increased few days before the storm and peak on the day of the storms, and then declined afterwards. HMFD values above $1.0 \times 10^{-6} \, \text{g kg}^{-1} \, \text{s}^{-1}$ is capable of producing substantial amount of rainfall mostly above 50 mm while wind shear has a much weaker impact on higher rainfall amount than moisture availability. CAPE above 1000 Jkg^{-1} and 1500 Jk^{-1} are favorable for convection over the southern and northern Nigeria, respectively. The study recommends quantitative analysis of moisture flux as a valuable short-term severe storm predictor and should be considered in the prediction of extreme rainfall.

Keywords

Mesoscale convective system · Moisture flux divergence · Wind shear · Extreme rainfall · Flood

Introduction

Human-induced climate change has increased the amount of water vapor in the atmosphere and has caused adverse effects on different regions, ecosystems, and economies across the world (Nwankwoala 2015). These effects depend not only on the sensitivity of populace to climate change but also on their ability to adapt to risks and changes associated with it. Since the atmospheric moisture budget plays an important role in the hydrology of a particular region, the changing weather patterns caused by climate change has increased the incidences of extreme rainfall events (Roshani et al. 2012; Weisman et al. 1988). Rainfall variability associated with climate change has impacted socioeconomic activities such as agriculture, food security, water resources management, health sector, hydroelectric power generation, and dam management among others in Nigeria (Bello 2010). Most of the negative effects of rainfall variability are on agriculture since majority of farmers in the country depend on rain-fed agriculture for livelihood (IPCC 2014; Lawal et al. 2016). Levels of adaptation of farmers in Nigeria to climate change is low, due to lack of adequate education, assets, information, and income (Madu 2016), and consequently agriculture is more vulnerable to climate change impact. Rainfall over Nigeria is mostly from the West African Monsoon systems (WAMS) (Diatta and Fink 2014), and agriculture is an important sector of economy of the country which is highly dependent on the WAMS (Raj et al. 2019). Studies have shown that the complexity of the atmospheric dynamics that generate rainfall, temporal and spatial variation of its scale made it difficult to understand and model. Also, the required parameters to predict it are usually complex even for a short-term period (Sumi et al. 2012). Majority of the results of studies conducted on extreme precipitation events over Nigeria showed that there have been some notable increase in intensity of rainfall extremes which usually claim many lives and properties (Okorie 2015).

In recent times, incidences of large storms have become more frequent with increased intensity, especially the occurrences of high rainfall in form of intense single-day events causing devastating flood (Enete 2014). However, short-range forecasting of these flood occurrences has been a great challenge. Studies have also shown that the distinctive property of West African monsoon flow is that there is seasonal, monthly, and daily variability in its moisture content mainly in the lowest 1 km of the atmosphere (Omotosho and Abiodun 2007). Furthermore, the intensity and duration of extreme rainfall are majorly dependent on the adequacy of moisture carried by the moist southwesterly flow from the South Atlantic Ocean. Studies have also shown that the daily variability of moisture advection mechanisms is responsible for the changes in the intensity and amount of rainfall (Omotosho et al. 2000). Couvreux et al. (2010) noted that at low level, daily moisture transport takes place with periodic northward advection of moisture flux that has 3–5 days frequency. A study conducted by Bechtold et al. (2004) also showed that large-scale thunderstorms in form of Mesoscale Convective Systems (MCSs) are formed over West Africa when there is constant supply of low-level moisture. The action of synoptic scale intensification of the St. Helena high pressure system over the South Atlantic Ocean is majorly responsible for moisture divergence into West Africa. Similarly, in a numerical study of moisture build-up and rainfall done by Omotosho and Abiodun (2007), it was reported that the rainfall amount does not depend on the monsoon flow alone but majorly on the sufficiency and the variability of its moisture content. Other studies have also examined the influence and interaction of different scales of motion such as African Easterly Jet (AEJ), Tropical Easterly Jet (TEJ), and African Easterly Waves (AEW) on the formation of MCSs (Nicholson 2013). These studies focused mainly on the scale interactions responsible for seasonal variability in weather pattern during northern summer (Janicot et al. 2011). They, however, did not consider in detail the quantity of daily moisture advection responsible for these rainfall extremes. The main focus of this chapter is to contribute to the understanding of the impact of moisture flux on the rainfall amount. Hence, it is necessary to study extreme precipitation events and diagnose the signatures of the meteorological parameters peculiar to such events. Studying this may enhance the assessment of the manner in which extreme rainfall events evolve and therefore provide a short-term early warning method to forecasters. It will also assist in understanding the evolution of some derived meteorological parameters such as the moisture flux which determines the quantity of rainfall and the wind shear which determines the life span of the storm (Weisman et al. 1988). Thorough understanding of these parameters will aid reliable short-term flood forecast. The usefulness of horizontal moisture flux at or near the earth's surface as a thunderstorm predictor has been recognized throughout various studies (Beckman 1990). Apart from the availability of moisture, sustenance of MCSs also requires a certain magnitude of vertical wind shear to produce a stronger and longer-lived system (Weisman and Rotunno 2004). Observational and numerical studies have revealed that horizontal winds and their vertical structures have important impacts on convective development; to buttress this point, Omotosho (1987) noted that thunderstorms occur, most frequently, in association with low-level wind shears below the AEJ (surface to 700 hPa) ranging from -20 to -5 s^{-1} and for mid troposphere (700–400 hPa) in the range of 0 to $10s^{-1}$. Despite the importance of wind shear, its effect on MCS has not been treated explicitly over Nigeria.

Most climate prediction models do not perform well in prediction of extreme rainfall events over West Africa because of their low resolution (Nyakwada 2004), the nature of parameterization schemes employed in the model, scarcity of real-time data, and mostly due to the convective nature of West African rainfall, hence the forecast of extreme rainfall event is a major challenge to forecasters in West Africa. The aim of this chapter is therefore to determine the impact of moisture flux, vertical wind shear, and other derived meteorological parameters such as wind divergence and convective available potential energy (CAPE) on MCSs during extreme rainfall events over Nigeria. Therefore, the study investigates the spatiotemporal variability of moisture flux that feeds into MCSs and its impacts on the occurrences of high impact rainfall. The objective is to assess the threshold values of derived meteorological parameters responsible for isolated cases of extreme precipitation in order to enhance its predictability and contribute to the understanding of the impact of moisture flux on the amount of precipitation. This research makes use of a synergistic approach involving the moisture flux divergence, wind shear analysis below and above the AEJ, and CAPE. The spatial distribution of these derived parameters is considered, with a focus on understanding their contribution to the formation and sustenance of MCSs during extreme rainfall events. This chapter will be a useful guide for further investigations into accurate prediction of high impact rainfall that can result into flood events using moisture flux analysis.

Description of the Study Area and Methodology

Study Area

Nigeria is situated between latitudes 4° and 14°N and longitudes 2° and 15°E and falls within the tropics. It shares borders with Niger in the north, Chad in the northeast, Benin in the west, Cameroon in the east, and its coast in the south borders the Gulf of Guinea on the South Atlantic Ocean. Precipitation is received mainly

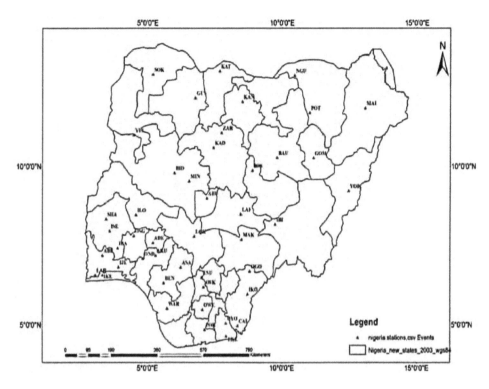

Fig. 1 Map of Nigeria showing the whole country as study area and spatial distribution of Nigerian Meteorological Agency synoptic stations used in this chapter

during the northern hemispheric summer, which is referred to as the wet seasons. Moist southwesterly winds from the South Atlantic Ocean prevail during the summer while dry northeasterlies from the Sahara desert are dominant in the winter which is the dry season (Fig. 1).

The confluence zone between both wind systems is the Inter-tropical Discontinuity (ITD). The surface location of the ITD significantly accounts for rainfall interannual variability in the country (Nicholson 2009). The ITD fluctuates seasonally during the northern summer over West Africa and migrates northward from its winter position of 4°N to its northernmost position of about 22°N (Fig. 2). The amount of rainfall experienced by different areas depends on the position of the ITD. Most of the convective rainfall follows the south-north-south displacement of the ITD (Sultan and Janicot 2003).

Data

Daily data for selected days of heavy rainfall was obtained from the European Centre for Medium-Range Weather Forecast's (ECMWF) ERA-INTERIM dataset on a gridded point of 0.125° × 0.125° and pressure levels of 1000, 850, 700, 400, and 200 hPa for the summer months of June–September 2010–2014. The daily datasets

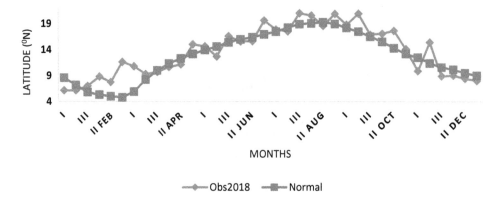

Fig. 2 Decadal latitudinal positions of the ITD in 2018 and its climatological mean over Nigeria. (Source: NiMet Climate Review Bulletin 2018)

are convective available potential energy (CAPE), specific humidity, zonal (U), meridional (V) winds, and divergence. Daily-accumulated rainfall was obtained from Tropical Rainfall Measuring Mission (TRMM) dataset, at a resolution of $0.25° \times 0.25°$. Observed rainfall for the months of June–September 2010–2014 was obtained from the Nigerian Meteorological Agency (NiMet) from 40 synoptic stations over Nigeria as shown in Fig. 1 to validate the TRMM dataset. Days of unusually high amount of rainfall during the months of June–September were selected from 2010 to 2014.

Research Methodology

Derived meteorological parameters such as moisture flux, CAPE, vertical wind shear, monsoon depth, the strengths of AEJ and TEJ were evaluated on these specific days to ascertain the characteristics of atmospheric dynamics during the occurrences of extreme rainfall events. The monsoon depth, the strengths of AEJ and TEJ are obtained by plotting the vertical wind profile using the U component of wind. The atmospheric dynamics of these events were diagnosed to find out the significant threshold of moisture flux, CAPE, vertical wind shear, monsoon depth, the strengths of AEJ and TEJ that may possibly be responsible for such unusually high amount of rainfall. Convective days were compared to a non-convective day to determine the difference in the characteristics of the atmospheric dynamics. Observationally, while there were occurrences of thunderstorms with heavy rainfall on convective days, none occurred on non-convective days. Ten weather events that produce rainfall above 50 mm were also selected for four meteorological stations over the northern region from 2010 to 2014. The mean of derived parameters such as moisture flux and CAPE 3 days prior and after the rainfall events are calculated to assess their characteristics during the period. (12.00°N, 8.59°E), Maiduguri (11.83 N, 13.15°E), Sokoto (13.01° N, 5.25°E), and Yelwa (10.83°N, 4.74°E) were chosen, because according to Omotosho (1985), about 90% of rainfall over these stations is attributed to MCSs.

Computation of Horizontal Moisture Flux and Wind Shear

Horizontal moisture flux is computed by the following formula.

Using U and V components of wind,

$$\nabla.V = \frac{\partial u}{\partial x} + \frac{\partial v}{\partial y} \tag{1}$$

where $\nabla. V$ = Divergence

Horizontal moisture flux divergence (HMFD) fields were derived from the specific humidity and components of the wind for the surface and 850 hPa by using the following formula:

$$\text{HMF} = -\left(U\frac{\partial q}{\partial x} + V\frac{\partial q}{\partial y} \right) + q\left(\frac{\partial u}{\partial x} + \frac{\partial v}{\partial y} \right) \tag{2}$$

where:

$$\left(U\frac{\partial q}{\partial x} + V\frac{\partial q}{\partial y} \right) \text{ is the Advection term}$$

$$\left(q\frac{\partial u}{\partial x} + \frac{\partial v}{\partial y} \right) \text{ is the divergence term}$$

where (q) is the specific humidity, (u) and (v) are zonal and meridional wind speed components of q, u and v; the advection term represents the horizontal advection of specific humidity, while the divergence term denotes the product of the specific humidity and horizontal mass convergence. $\frac{\partial}{\partial x}$ and $\frac{\partial}{\partial y}$ show the horizontal variation of atmospheric quantities such as specific humidity and wind. The first term in the moisture flux (MF) equation is moisture advection. This term incorporates changes of the moisture field with time or the flux of the moisture field. The moisture advection term, similar to the mass divergence term, incorporates into MF the effects of moisture availability on convection. The second term, mass convergence term incorporates moisture accumulation by multiplying the wind convergence, which is the rate at which the air itself is pooling, by the moisture content of the air (mixing ratio). The mass convergence term is usually the dominant term, and observations have shown that the moisture advection term can significantly contribute to the development and subsequent intensification of storms. By combining the effects of mass convergence as a low-level forcing mechanism with moisture availability, moisture flux incorporates most of the ingredients necessary for convection (Roshani et al. 2012).

Vertical wind shear (U_s) was calculated using the zonal component of wind at 1000, 700, 400, and 200 hPa levels. The vertical wind shear (U_s) is defined as:

$$U_s = \frac{du}{dz}S^{-1} \tag{3}$$

The vertical wind shear at the lower, mid, and upper troposphere using the formula (Omotosho 1987):

$$U_{S(L)} = U_{700} - U_{surface} \tag{4}$$

$$U_{S(M)} = U_{400} - U_{700} \tag{5}$$

$$U_{S(U)} = U_{200} - U_{400} \tag{6}$$

where:

$U_{S(L)}$ = Wind shear at lower troposphere
$U_{S(M)}$ = Wind shear at mid troposphere
$U_{S(U)}$ = Wind shear at upper troposphere

Rainfall events that produced rainfall above 50 mm were also selected for four meteorological stations over the northern Nigerian region from 2010 to 2014. The mean of derived moisture flux and CAPE 3 days prior and after the rainfall events were calculated to assess their characteristics during the period. Kano (12.00°N, 8.59°E), Maiduguri (11.83°N, 13.15°E), Sokoto (13.01°N, 5.25°E), and Yelwa (10.83°N, 4.74°E) were chosen, because according to Omotosho (1985), about 90% of rainfall over these stations is attributed to MCSs.

Result

Convective Days

Moisture Flux Analysis

The study presents the analysis of rainfall events that took place from the first to third July 2014 over most parts of the country. The intense rainfall observed led to flooding which resulted into loss of lives and properties. The TRMM rainfall analysis over the country from first to third as shown by Fig. 3a–c depicted the widespread rainfall across the country during the period. Rainfall amount of 71.5, 104, and 117 mm was observed on the first to third day, respectively, over Eket (4.65°N, 7.94°E), a coastal city in the southeastern part of the country. The horizontal moisture flux divergence (HMFD) analysis indicated that divergence of moisture at the surface was from the South Atlantic Ocean for the 3 days considered. Figure 4a–c showed that the values of HMFD ranged from 0.10 to 1.05×10^{-6} g kg^{-1} s^{-1} on the first day with slight reduction on the second day to 0.85×10^{-6} g kg^{-1} s^{-1} while the maximum value of HMFD on the third day was up to 1.25×10^{-6} g kg^{-1} s^{-1}. This analysis showed that higher the amount of moisture supplied to the storm, the higher the amount of precipitation. The analysis at 850 hPa level as shown by Fig. 5a–c depicted that area of moisture divergence shifted from the coast to mainly over the high grounds of Jos, Mambilla plateaus, and other high

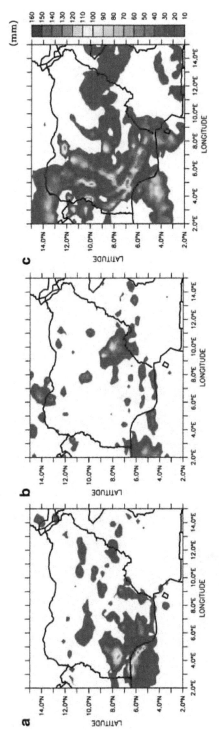

Fig. 3 (**a–c**) Spatial distribution of TRMM rainfall (color; mm) for first, second, and third of July 2014, respectively, over Nigeria

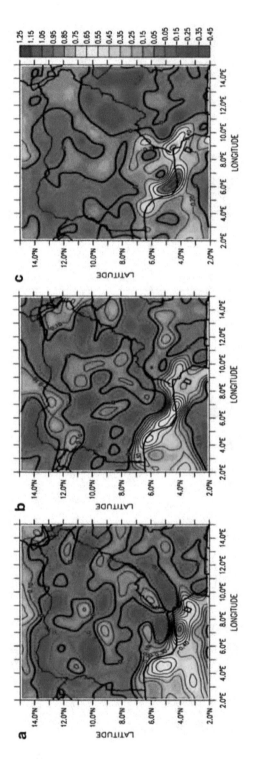

Fig. 4 (a–c) Spatial distribution of horizontal moisture flux divergence (HMFD) at the surface (1000 hPa level) for 1–3 July 2014 (conversely, dotted lines indicate moisture convergence)

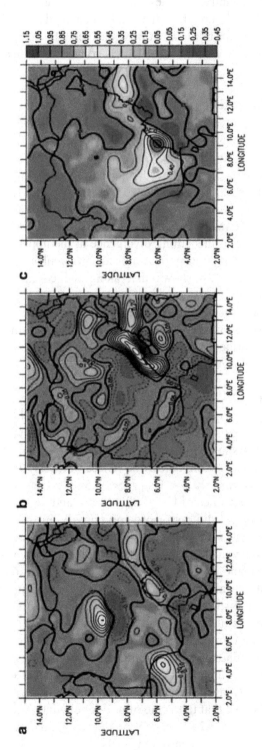

Fig. 5 (**a–c**) Spatial distribution of horizontal moisture flux divergence (HMFD) at the 850 hPa for 1–3 July 2014 (conversely, dotted lines indicate moisture convergence)

grounds across the country. Values of HMFD ranged from 0.01 to 0.55×10^{-6}g kg^{-1} s^{-1} on the first and second day and increased to 1.15×10^{-6}g kg^{-1} s^{-1} on the third day. It was observed that close to the divergence areas were areas of strong convergence, which coincided with areas of highest precipitation.

Wind Divergence Analysis

Figures 6a–c and 7a–c showed the observed rainfall distribution over the country was maximum along the coastal areas and some inland cities of the southwest, e.g., at Shaki (8.35°N, 5.47°E) on the first day, strong wind divergence at 200 hPa level exactly over this area enhanced the lower tropospheric wind convergence. This is consistent with Nicholson (2009) which stated that strong upper-level divergence is associated with strong upward motion and severe convective storms; in contrast, upper-level convergence usually indicates downward motion, which is a sign of decaying convection. Due to this strong divergence at the upper level, convection became vigorous and vertical transport of moisture was enhanced from the lower troposphere. However, Fig. 5b shows that along the coast, convergence was observed at the surface with corresponding divergence at 200 hPa with little amount of rainfall; this may also be attributed to insufficient moisture as shown by the moisture flux analysis. Similarly, over the western parts of the country, wind divergence was at the surface with corresponding convergence at 200 hPa, hence less precipitation was observed.

Vertical Wind Shear

The wind shear analysis, Fig. 8a–c showed that the low-level wind shear Us$_{(L)}$ ranged between -4 s^{-1} and -14 s^{-1} across the country. These values agreed with Omotosho (1987) who showed that thunderstorms occur most frequently in association with low-level shears, below the African Easterly Jet (i.e., surface to 700 hPa) with values within $-20 \sim < $ Us$_{(L)} \sim < -5$ s^{-1}. Areas with precipitation value above 50 mm have values of Us $_{(L)}$ of -14 s^{-1} and above. However, some areas with Us $_{(L)}$ of -14 s^{-1} did not record any rainfall, this may be attributed to wind divergence observed at the surface or inadequate moisture supply (Grist and Nicholson, 2001). The value of Us $_{(L)}$ ranges from -8 to -4 s^{-1} over the southwest on the first and second day and up to -12 s^{-1} on the third day. Considerable amount of rainfall observed over these regions coincided with areas with adequate moisture flux convergence on the first and third day. According to Rotunno et al. (1988) and Weisman et al. (1988), vertical wind shear is important in the formation of organized long-lived convection; however, very strong horizontal wind shear can inhibit the growth of cumulus clouds by blowing away the parts of the cloud containing the best developed precipitation particle and thereby preventing the process of precipitation (Rickenbach et al. 2002). Figure 8a–c showed that the value of mid-level wind shear Us $_{(U)}$ over the country ranges between 0 and 2 s^{-1} except on the second day where the value of W$_S$U over western parts is between 0 and -10s^{-1}. It is noteworthy that moderately sheared environment is important for sustaining MCSs during extreme rainfall event (Figs. 9 and 10).

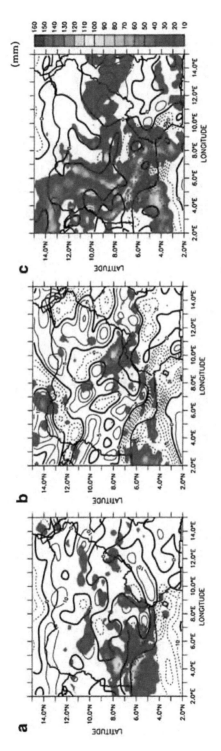

Fig. 6 (a–c) Spatial distribution of rainfall (color; mm) and wind divergence at surface (contour; s⁻¹) for 1–3 July 2014 (dotted lines indicate convergence)

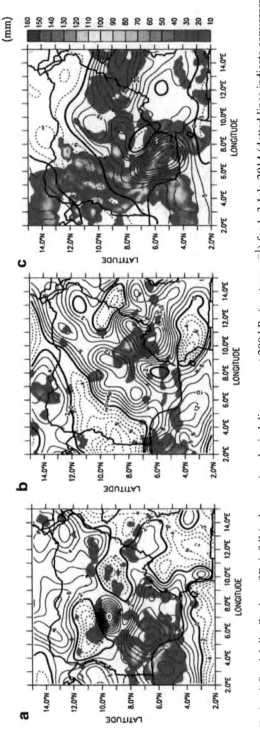

Fig. 7 (a–c) Spatial distribution of Rainfall (color; mm) and wind divergence at 200 hPa (contour; s⁻¹) for 1–3 July 2014 (dotted lines indicate convergence)

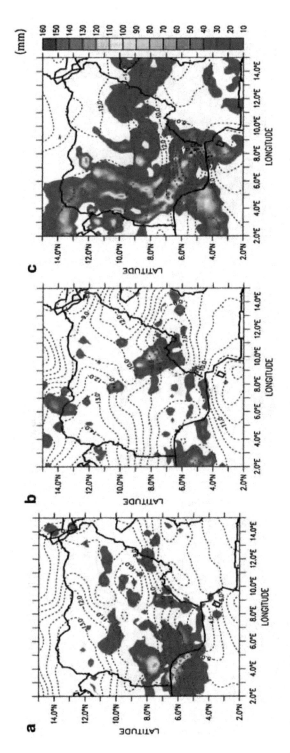

Fig. 8 (a–c) Spatial distribution of low-level wind shear $U_{S(L)}$ between the surface (1000 hPa) and 700 hPa level (contour; s^{-1}) and TRMM precipitation (color; mm) for 1–3 July 2014

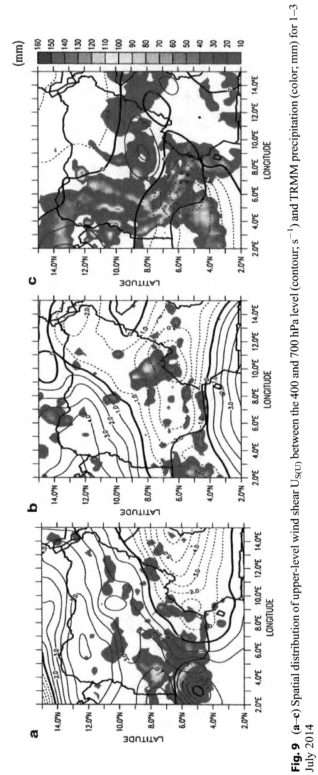

Fig. 9 (a–c) Spatial distribution of upper-level wind shear $U_{S(U)}$ between the 400 and 700 hPa level (contour; s^{-1}) and TRMM precipitation (color; mm) for 1–3 July 2014

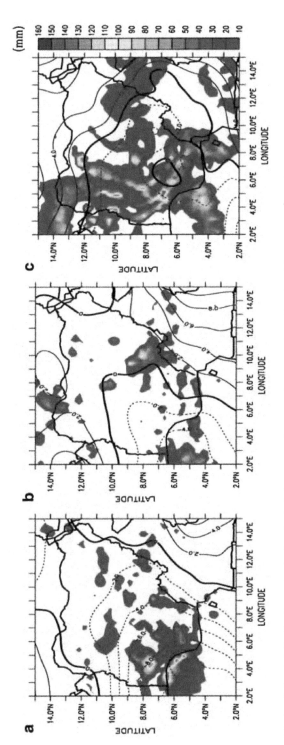

Fig. 10 (**a**–**c**) Spatial distribution of upper-level wind shear $U_{S(U)}$ between the 200 and 400 hPa level (contour; s^{-1}) and TRMM precipitation (color; mm) for 1–3 July 2014

Wind Vector at 850 hPa

Figure 11a–c showed that there was continuous moisture supply from the Atlantic Ocean throughout the rainfall events. Moist southwesterly winds convergence that was observed along the southwest coast on the first day produced significant amount of rainfall over Lagos (6.52°N, 3.37°E) and Lokoja (7.80°N, 6.73°E) axis. The wind flow from south to north remained steady from the day 1 and 2. Although over the northern parts, the wind direction indicated northeasterly flow on day 3, but sufficient residual moisture has already accumulated over the country up to the northern areas before the occurrence of a more widespread and heavy rainfall on the third day as shown by Fig. (11a–b). On the third day, over the southeastern parts, there was a well-organized deep monsoon flow from the Gulf of Guinea feeding into a vortex over the inland area of the southeastern parts of the country. The advected moisture depth was enough to maintain the active system over the southeastern axis as shown by the HMFD analysis. A total rainfall of 292.5 mm was recorded round Eket (4.65° N, 7.94°E), the vicinity of the vortex.

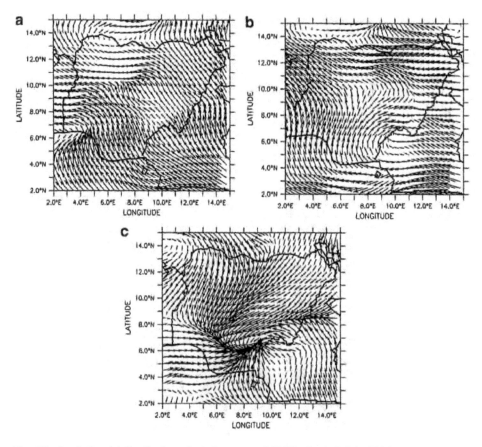

Fig. 11 (a–c): Spatial distribution of wind vector at 850 hPa for 1–3 July 2014

Convective Available Potential Energy (CAPE)

CAPE is an energy-based measure of atmospheric stability; it is a very important index in the forecasting of rainfall over Nigeria (Olaniyan et al. 2015). Figure 12a–c showed that the CAPE up to 2900 Jkg^{-1} was observed over the northeast and central parts in the first and second day; however, not much precipitation was recorded. This may also be attributed to insufficient moisture as shown by Fig. 4a–b. Over the southwest, CAPE value of 1600 Jkg^{-1} and 2400 Jkg^{-1} was observed on the first day and the second day, respectively, but due to reduction in moisture flux on the second day, rainfall reduction was observed compared to the first day. Over the southeastern parts, the CAPE values vary from 800 JKg^{-1} to 1600 JKg^{-1} and 1000–1500 Jkg^{-1} on the first and second day, respectively, while on the third day, the value ranged from 1000 JKg^{-1} to 1500 JKg^{-1}, and the highest amount of rainfall was observed on the third day. Over the northern parts, CAPE values favorable for convection ranges between 1000 and 2800 Jkg^{-1} on the first day, while it is 600–3000 Jkg^{-1} on the second day though no rainfall was observed on this day due to reduced moisture flux. On the third day, the CAPE value was 2500 Jkg^{-1} and peaked to 4000 Jkg^{-1}, higher rainfall was observed due to abundant moisture over this area indicating a deep layer of moisture to fuel the MCSs. A consistent pattern of CAPE was observed throughout the rainfall events, CAPE increases from coastal to the northern parts of Nigeria and the higher the CAPE, the higher the intensity of storm, provided moisture is sufficient.

Vertical Wind Profile (AEJ, TEJ, and Monsoon Depth)

Figure 13a–c showed the extent of the moist southwesterly winds, the strength of the AEJ and the TEJ over the south and northern regions, and their mean position over the whole country, respectively, from the first to third of July. The zonal wind at the surface was westerly with speed of about 2, 6, and 5 ms^{-1}, respectively, over the south of 9°N, north of 9°N, and the entire country. The first and second days indicated lower moisture depth at 900 hPa compared to 950 hPa on the third day (Fig. 13c). The AEJ was located at about 700 hPa with a mean speed of 12, 10, and 8 ms^{-1} for the first, second, and third day, respectively, while the speed of TEJ was 18, 12, and 13 ms^{-1}, respectively, for the 3 days over the entire country. Thermodynamically, the AEJ is often responsible for advection of both sensible heat and latent energy into regions where severe thunderstorms are formed. TEJ is responsible for enhancing upper level divergence, which in turn encourages vertical motion and lower level convergence (Nicholson et al. 2012). Therefore, the strength of the jets and the availability of adequate moisture support these extreme rainfall events.

Non-convective Day

Generally, the atmosphere was stable on this day; most stations in the country reported no rainfall. Analysis of a non-convective day was done to evaluate the

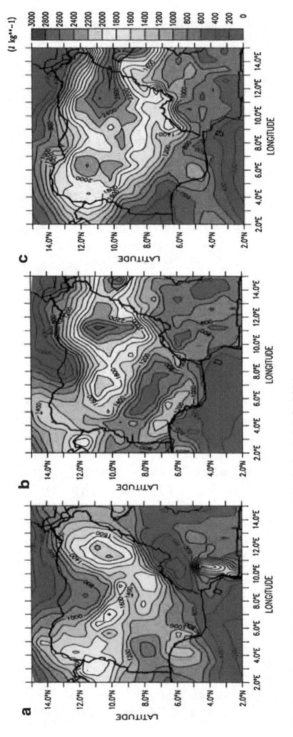

Fig. 12 (**a–c**): Spatial distribution of CAPE over Nigeria for 1–3 July 2014

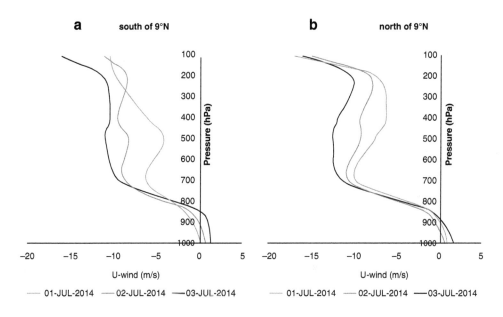

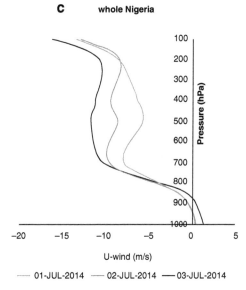

Fig. 13 (**a–c**) Vertical wind profiles (in m/s) averaged over (**a**) south of 9°N, (**b**) north of 9°N, and (**c**) the entire Nigeria for 1–3 July 2014. Days 1, 2, and 3 are indicated in blue, red, and gray lines, respectively

difference in the behavior of the convective parameters in order to identify the reason for the observed stability in the atmosphere on the day.

Divergence Analysis

Figure 14a, b shows that over the southwestern parts of the country, convergence was observed at the surface and also at 200 hPa, and hence vertical motion was

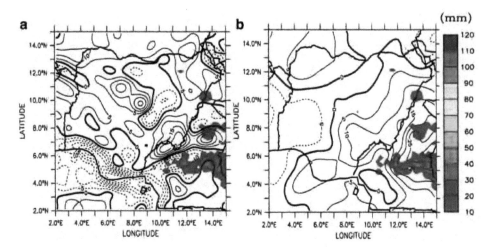

Fig. 14 Spatial distribution of rainfall (color; mm) and wind divergence at (**a**) 1000 hPa (**b**) 200 hPa level (contour; s^{-1}) for 27 August 2014 (dotted lines indicate convergence)

suppressed; this may be due to widespread subsidence prominent during this period caused by reduced sea surface temperature and the ridging effect of the south Atlantic high pressure system over the St. Helena. This period is usually referred to as the little dry season. However, over some parts of the central region and the southeastern coast, convergence was observed at the surface while there was corresponding divergence at 200 hPa, but no rainfall was observed; this may be due to lack of sufficient moisture as shown by the HMFD (as shown in the next section) analysis. The rest of the country was prevailed by convergence at 200 hPa while divergence was observed at the surface. These features could suppress vertical transport of moisture.

Moisture Flux Analysis

The HMFD analysis at the surface in Fig. 15a shows that less moisture was available at the surface and 850 hpa level. Moisture flux diverging from the Atlantic Ocean at the surface over the southwest coast was almost negligible, while over the southeastern part extended from the coast even to the inland has values ranging from 0.10 to 0.55×10^{-6} kg/gs^{-1}. Although wind convergence is present at the surface with corresponding divergence at 200 hPa, moisture may not be sufficient as shown in Fig. 14b, hence no rainfall is observed over the country.

Vertical Wind Shear Analysis

Figure 16a–c shows that $U_{S(L)}$ values ranged between 0 s^{-1} and -12 s^{-1}. The value of $U_{S(M)}$ across the country ranged from 0 s^{-1} to -8 s^{-1}, zero value was observed over the central state, while the value of $U_{S(U)}$ ranged from -8 s^{-1} to -18 s^{-1} as shown in Fig. 15a–c. Although this range is conducive for storm initiation, provided other conditions such as moisture availability, lower level convergence, and upper level divergence are met; otherwise, convection may be suppressed.

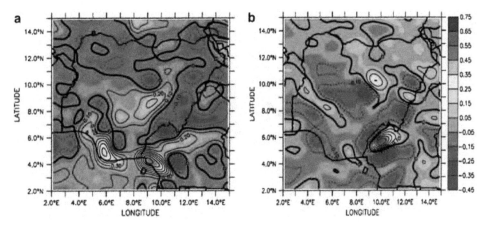

Fig. 15 (**a** and **b**) Spatial distribution of horizontal moisture flux divergence (HMFD) at the surface (1000 hPa) and 850 hPa for 27 August 2014 (conversely, dotted lines indicate moisture convergence)

Wind Vector

The wind vector analysis shown in Fig. 17 depicted the prevalence of southwesterly wind from the Atlantic Ocean, but no rainfall was over the country; this was confirmed by insufficient moisture as shown in the moisture flux analysis of Fig. 15a, b despite the fact that the entire country was prevailed by southwesterly winds, and no rainfall was recorded across the country. According to Omotosho and Abiodun (2007), little or no precipitation is observed below a certain limit of atmospheric moisture.

Convective Available Potential Energy (CAPE)

Figure 18 shows that the CAPE across the country ranges between 200 Jkg^{-1} in the south and 1200 Jkg^{-1} over the north; this is an indication of less potential energy to support convection, and for this particular day, moisture was also not sufficient to support rainfall as shown by HMFD analysis (Fig. 15a, b).

Mean HMFD and CAPE at the Surface, Three Days Prior and Three Days After Storm Events

Ten weather events that produce rainfall above 50 mm were selected for four meteorological stations over the northern Nigeria from 2010 to 2014. The mean of derived parameters for selected days namely, horizontal moisture flux divergence (HMFD) and CAPE were evaluated 3 days before and 3 days after the storm over the northern stations of Kano (12.00°N, 8.59°E), Maiduguri (11.83 N, 13.15°E), Sokoto (13.01°N, 5.25°E), and Yelwa (10.83°N, 4.74°E) depicted in Figs. 19a, b, 20a, b, 21a, b, and 22a, b, respectively. On these figures, horizontal axes represent days; with 0 representing the day of the stormy events while negative and positive values (day) represent, respectively, 3 days prior and 3 days after the storm. Generally,

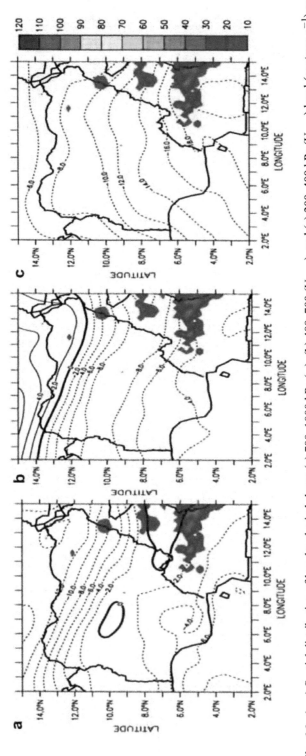

Fig. 16 (**a**–**c**) Spatial distribution of low-level wind shear: (**a**) 700–1000 hPa ($_L$), (**b**) 400–700 ($U_{S(M)}$), and (**c**) 200–400 hPa ($U_{S(U)}$) level (contour; s^{-1}) and TRMM precipitation (color; mm) for 27 August 2014

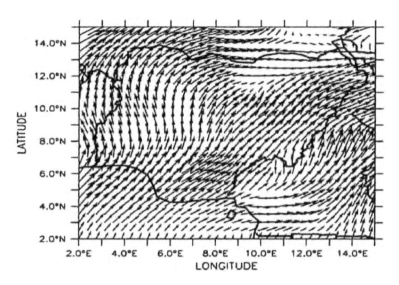

Fig. 17 Spatial distribution of wind vector over Nigeria at 850 hPa for 27 August 2014

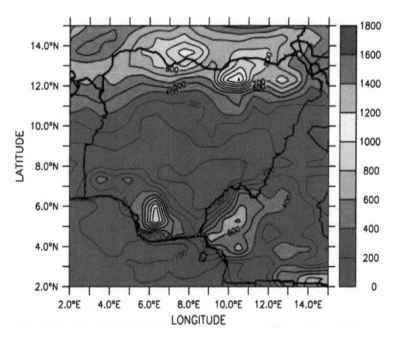

Fig. 18 Spatial distribution of CAPE over Nigeria for 27 August 2014

moisture divergence from the Atlantic Ocean accumulates prior to the storm, reaches climax on the day of the rainfall event, and starts declining after the storm (Panel (a) of Figs. 19, 20, 21, and 22). The mean HMFD was highest on day 0 which is day of the rainstorm. This is depicted by negative moisture flux and afterwards it gradually decreased after the storm. The analysis showed that extreme rainfall events are characterized by significant moisture flux divergence prior to storm events. Panel

(a)

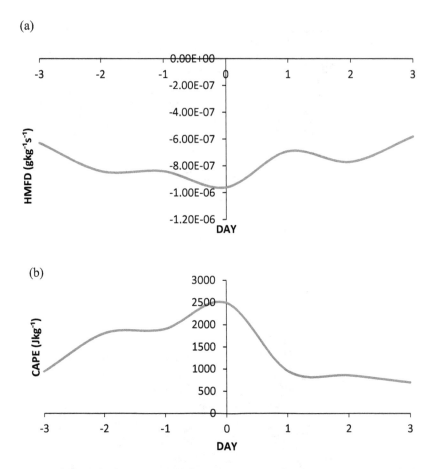

Fig. 19 (**a** and **b**) Derived meteorological parameters at surface over Kano (3 days before and 3 days after extreme rainfall events from June to September 2010–2014) for (**a**) mean HMFD and (**b**) CAPE

(b) of Figs. 19, 20, 21, and 22 also showed a similar pattern for the mean CAPE analysis.

Table 1 shows the mean CAPE on day 0, which ranges from 2416 Jkg^{-1} to 2954 Jkg^{-1} with the highest over Sokoto.

Conclusion

This chapter identified and examined a set of severe widespread rainfall that produced flood events over different parts of Nigeria. The result showed that moisture, convective instability, vertical wind shear, and lifting mechanisms all contributed to these events, but most importantly, the moisture influx. The quantity of rainfall over a given area can be related to the magnitude of the lower tropospheric moisture flux. The study also showed that the transfer of moisture flux in the low layer is mainly from the South Atlantic Ocean, and the higher the moisture flux diverged from the

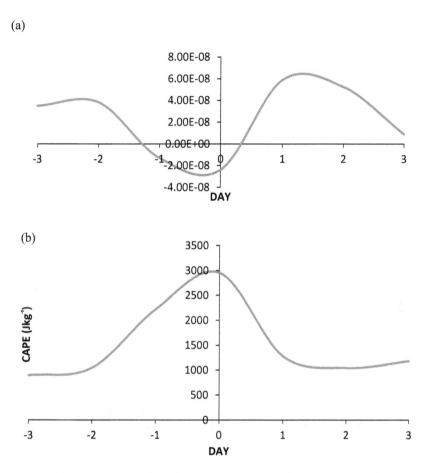

Fig. 20 (**a** and **b**) Derived meteorological parameters at surface Sokoto (3 days before and 3 days after the storm for extreme events from June to September 2010–2014) for (**a**) mean moisture flux divergence and (**b**) CAPE. Negative values in moisture flux analysis indicate convergence while positive values indicate divergence

Atlantic Ocean, the higher the amount of rainfall. The result also indicated that low-level convergence that corresponds with upper-level divergence encourages vertical transport of moisture while low-level divergence and upper-level convergence results in subsidence. At the surface, the value of moisture flux divergence ranges between 0.05 and 1.15×10^{-6}gkg^{-1} s^{-1} at the vicinity of areas where considerable amount of precipitation of above 50 mm were observed. They are mostly located westward of moisture divergence zone. This gives a good indication of where flood is most likely. Moisture flux divergence value ranged from 1.0 to 2.0×10^{-6} gkg^{-1}s^{-1} at 850 hpa around the areas with substantial amount of rainfall in its vicinity, the high grounds of the southwest, Jos, Mambila, Adamawa plateaus, and Cameroonian mountain are good source of moisture divergence at 850 hPa, and hence theses area can be referred to as fertile ground for convection (Hodges and Thorncroft 1997; Akinsanola and Ogunjobi 2014). The wind shear below the AEJ ($U_{S(L)}$) over

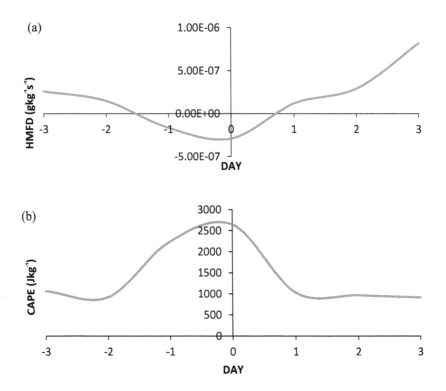

Fig. 21 (**a** and **b**) Derived meteorological parameters at surface over Maiduguri (3 days before and 3 days after extreme rainfall events from June to September 2010–2014) for (**a**) mean moisture flux divergence and (**b**) CAPE. Negative values in moisture flux analysis indicate convergence while positive values indicate divergence

the region of intense precipitation ranges between -8 and -12 ms^{-1}. At mid-troposphere ($U_{S(M)}$), wind shear value ranged from 2 s^{-1} to -8 s^{-1}, while at upper level ($U_{S(U)}$), the values ranged between 0 and 12 s^{-1}. Most of the result of $U_{S(L)}$ were in agreement with Omotosho (1987) on the value of wind shear necessary for the initiation and sustenance of MCSs.

The CAPE analysis indicated that potential energy equal or greater than 1500 Jkg^{-1} favored convection over the northern parts, while CAPE value equal or greater 1000 Jkg^{-1} was able to trigger convective activities over the southern parts. The result also showed that extreme rainfall also depends on convective available potential energy CAPE, and the higher the CAPE, the more intense was the rainfall, provided moisture was sufficient. Similarly, rainfall observed at a particular area varies according to the amount of moisture flux advected by the monsoon winds into the area. The mean moisture flux and CAPE analysis for extreme storm events over selected cities in northern Nigeria indicated that there is always an increase in the value of moisture flux and CAPE 3 days prior to storm occurrences. This can be a good indicator for forecasting extreme rainfall events. Table 1 shows the peak values of mean CAPE and moisture flux over the selected northern Nigerian stations. The result also shows that sufficient CAPE and wind

(a)

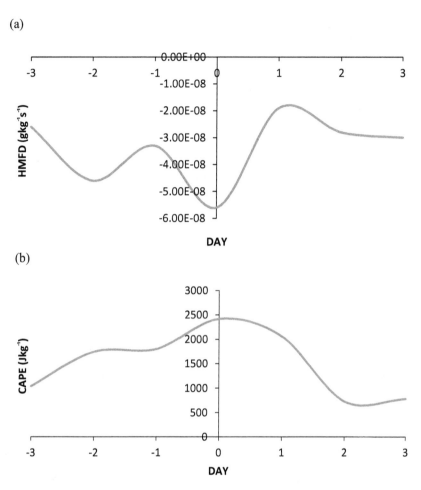

(b)

Fig. 22 (**a** and **b**) Derived meteorological parameters at surface over Yelwa (3 days before and 3 days after extreme rainfall events from June to September 2010–2014) for (**a**) mean moisture HMFD and (**b**) CAPE

Table 1 Peak values of mean derived meteorological parameters at the surface HMFD and CAPE over Kano, Maiduguri, Sokoto, and Yelwa for JJAS (2010–2014)

STATIONS	HMFD (gKg^{-1} s^{-1})	CAPE (Jkg^{-1})
KANO	−9.60E-07	2492
MAIDUGURI	−2.90E-07	2644
SOKOTO	−2.40E-08	2954
YELWA	−5.60E-08	2416

shear is not enough for convection, it is necessary to have sufficient amount of moisture for initiation and sustenance of the storm throughout its life. These observed pattern of wind flow at the surface and 850 hPa, the CAPE and HMFD when sighted on the forecast charts may be good indicators for forecasting extreme rainfall and the likelihood of flood events which will help in early preparedness and prevention of the worst impacts of such extreme events on lives and properties.

A further study of moisture flux and wind shear is recommended using different operational models and more network of stations as this will give a better perception of impact of moisture flux on spatial rainfall variability across the country. Forecasting precipitation amount is a challenging task for forecasters, therefore adequate study of the criteria such as moisture flux, wind shear, and CAPE will increase the understanding of extreme rainfall events; though there will always be variability in the values of meteorological parameters, sound understanding of forecast models, learning how to analyze the situation using the appropriate tools, and knowing how to apply these tools will give the best chance of predicting extreme events and issuing timely warnings.

References

Akinsanola AA, Ogunjobi KO (2014) Analysis of rainfall and temperature variability over Nigeria. Glob J Hum-Soc Sci: B 14(3):11–17

Bechtold P, Chaboureau J-P, Beljaars A, Betts AK, Kohler "M, Miller M, Redelsperger JL (2004) The simulation of the diurnal cycle of convective precipitation over land in a global model. Q J Roy Meteor Soc 130:3119–3137

Beckman SK (1990) A study of 12 h NGM low-level moisture flux convergence centers and the location of severe thunderstorms/heavy rain. In: Proceedings of the 16th AMS conference on severe local storms, Kananaskis Park, Alta, pp 78–83

Bello NJ (2010) Impacts of climate change on food security in sub-Saharan Africa. In: Proceedings of the 14th Annual Symposium of the International Association of Research Scholars and Fellows, IITA, Ibadan, pp 13–25

Couvreux F, Guichard F, Bock O, Campistron B, Lafore JP, Redelsperger JL (2010) Synoptic variability of the monsoon flux over West Africa prior to the onset. Q J R Meteorol Soc 136 (1):159–173

Diatta S, Fink AH (2014) Statistical relationship between remote climate indices and West African monsoon variability. Int J Climatol 34:3348–3367

Enete IC (2014) Impacts of climate change on agricultural production in Enugu State, Nigeria. J Earth Sci Clim Change 5(9):234. https://www.omicsonline.org/open-access/impacts-of-climate-change-on-agricultural-production-in-enugu-state-nigeria-2157-7617.1000234.php?aid=32633

Grist JP, Nicholson SE (2001) A study of the dynamic factors influencing the rainfall variability in the West African Sahel. J Clim 14(7):1337–1359

Hodges KI, Thorncroft CD (1997) Distribution and statistics of African mesoscale convective weather systems based on the ISCCP Meteosat imagery. Mon Weather Rev 125:2821–2837. https://doi.org/10.1175/1520-0493(1997)125<2821:DASOAM>2.0.CO;2

IPCC (2014) Synthesis report. In: Core Writing Team, Pachauri RK, Meyer LA (eds) Contribution of working groups I, II and III to the fifth assessment report of the Intergovernmental Panel on Climate Change. IPCC, Geneva, 151pp

Janicot S, Caniaux G, Chauvin F, De Coëtlogon G, Fontaine B, Hall N (2011) Intraseasonal variability of the West African monsoon. Atmos Sci Lett 12(1):58–66

Lawal KA, Abatan AA, Anglil O, Olaniyan E, Olusoji Victoria H, Oguntunde PG, Lamptey B, Babatunde JA, Shiogama H, Michael FW, Dith AS (2016) The late onset of the 2015 wet season in Nigeria. BAMS 97:63–69. https://doi.org/10.1175/BAMSD-16-0131.1

Madu IA (2016) Rurality and climate change vulnerability in Nigeria: assessment towards evidence based even rural development policy. Paper presented at the 2016 Berlin conference on global environmental change, 23–24 May 2016 at Freie Universität Berlin. https://pdfs. semanticscholar.org/508b/94cab07b84a703b44eca1089326cc98d7495.pdf?_ga=2.154518008. 112403230.1572433568-162569160.1557482164

Nicholson SE (2009) A revised picture of the structure of the "monsoon" and land ITCZ over West Africa. Clim Dyn 32(7–8):1155–1171

Nicholson SE (2013) The West African Sahel: a review of recent studies on the rainfall regime and its interannual variability. Int Scholar Res Notices 2013:453521, 32 p. https://doi.org/10.1155/2013/453521

Nicholson SE, Klotter DA, Dezfuli AK (2012) Spatial reconstruction of semi-quantitative precipitation fields over Africa during the nineteenth century from documentary evidence and gauge data. Quat Res 78:12–23

NiMet Climate Review Bulletin (2018) Nigerian Meteorological Agency 2019

Nwankwoala HNL (2015) Causes of climate and environmental changes: the need for environmental-friendly education policy in Nigeria. J Educ Pract 6(30). http://www.iiste.org. ISSN 2222–1735 (Paper) ISSN 2222-288X (Online)

Nyakwada W (2004) The challenges of forecasting severe weather and extreme climate events in Africa WMO WSHOP-SEEF/Doc.4(4)

Okorie FC (2015) Analysis of 30 years rainfall variability in Imo state of South-Eastern Nigeria. In: Hydrological sciences and water security: past, present and future. IAHS Press, Wallingford, pp 131–132

Olaniyan E, Afiesimama EOF, Lawal KA (2015) Simulating the daily evolution of West African monsoon using high resolution regional Cosmo-model: a case study of the first half of 2015 over Nigeria. J Climatol Weather Forecast 33:3–8

Omotosho JB (1985) The separate contributions of squall lines, thunderstorms and the monsoon to the total rainfall in Nigeria. J Climatol 5:543–552

Omotosho JB (1987) Richardson number, vertical wind shear, and storm occurrences over Kano Nigeria. Atmos Res 21:123–137

Omotosho JB, Abiodun BJ (2007) A numerical study of moisture build-up and rainfall over West Africa. Meterol Appl 14:209–225

Omotosho JB, Balogun AA, Ogunjobi K (2000) Predicting monthly and seasonal rainfall, onset and cessation of the rainy season in West Africa using only surface data. Int J Climatol 20:865–880

Raj J, Bangalath HK, Stenchikov G (2019) West African monsoon: current state and future projections in a high-resolution AGCM. Clim Dyn 52:6441–6461. https://doi.org/10.1007/s00382-018-4522-7

Rickenbach TM, Ferreira RN, Halverson J, Silva Dias MAF (2002) Mesoscale properties of convection in Western Amazonia in the context of large-scale wind regimes. J Geophys Res Atmos 107:8040

Roshani A, Fateme P, Zahra H, Hooshang G (2012) Studying the moisture flux over South and Southwest of Iran: a case study from December 10 to 13, 1995 rain storm. Earth Sci Res 2 (2):2013

Rotunno R, Klemp JB, Weisman ML (1988) A theory for strong long lived squall lines. National Center for Atmospheric Research Boulder, Colarado, pp 463–483

Sultan B, Janicot S (2003) The West African monsoon dynamics. Part II: the "preonset" and "onset" of the summer monsoon. J Clim 16:3407–3427

Sumi SM, Zaman MF, Hirose H (2012) A rainfall forecasting method using machine learning models and its application to the Fukuoka City case. Int J Appl Math Comput Sci 22(4):841–854

Weisman ML, Rotunno R (2004) A theory for strong long-lived squall lines revisited. J Atmos Sci 61:361–382. https://doi.org/10.1175/1520-0469(2004)061<0361:ATFSLS>2.0.CO;2

Weisman ML, Klemp JB, Rotunno R (1988) Structure and evolution of numerically simulated squall lines. J Atmos Sci 45:1990–2013

PERMISSIONS

The contributors of this book come from diverse backgrounds, making this book a truly international effort. This book will bring forth new frontiers with its revolutionizing research information and detailed analysis of the nascent developments around the world.

We would like to thank all the contributing authors for lending their expertise to make the book truly unique. They have played a crucial role in the development of this book. Without their invaluable contributions this book wouldn't have been possible. They have made vital efforts to compile up to date information on the varied aspects of this subject to make this book a valuable addition to the collection of many professionals and students.

This book was conceptualized with the vision of imparting up-to-date information and advanced data in this field. To ensure the same, a matchless editorial board was set up. Every individual on the board went through rigorous rounds of assessment to prove their worth. After which they invested a large part of their time researching and compiling the most relevant data for our readers.

The editorial board has been involved in producing this book since its inception. They have spent rigorous hours researching and exploring the diverse topics which have resulted in the successful publishing of this book. They have passed on their knowledge of decades through this book. To expedite this challenging task, the publisher supported the team at every step. A small team of assistant editors was also appointed to further simplify the editing procedure and attain best results for the readers.

Apart from the editorial board, the designing team has also invested a significant amount of their time in understanding the subject and creating the most relevant covers. They scrutinized every image to scout for the most suitable representation of the subject and create an appropriate cover for the book.

The publishing team has been an ardent support to the editorial, designing and production team. Their endless efforts to recruit the best for this project, has resulted in the accomplishment of this book. They are a veteran in the field of academics and their pool of knowledge is as vast as their experience in printing. Their expertise and guidance has proved useful at every step. Their uncompromising quality standards have made this book an exceptional effort. Their encouragement from time to time has been an inspiration for everyone.

The publisher and the editorial board hope that this book will prove to be a valuable piece of knowledge for researchers, students, practitioners and scholars across the globe.

LIST OF CONTRIBUTORS

Nnyaladzi Batisani
Botswana Institute for Technology Research and Innovation, Gaborone, Botswana
Food and Agriculture Organization of the United Nations, Rome, Italy

Flora Pule-Meulenberg, Utlwang Batlang and Nelson Tselaesele
Botswana University of Agriculture and Natural Resources, Gaborone, Botswana

Federica Matteoli
Food and Agriculture Organization of the United Nations, Rome, Italy

Mary Nthambi
Department of Environmental Economics, Brandenburg University of Technology Cottbus-Senftenberg, Cottbus, Germany

Uche Dickson Ijioma
Department of Raw Material and Natural Resource Management, Brandenburg University of Technology Cottbus-Senftenberg, Cottbus, Germany

Bernhard Freyer
Division of Organic Farming, University of Natural Resources and Life Sciences (BOKU), Vienna, Austria

Jim Bingen
Michigan State University (MSU), East Lansing, MI, USA

Oluwole Olutola
University of Johannesburg, Johannesburg, South Africa

Rebecca Sarku
University for Development Studies, Tamale, Ghana

Divine Odame Appiah
Environmental Management Practice Research Unit, Department of Geography and Rural Development, Faculty of Social Sciences, Kwame Nkrumah University of Science and Technology, Kumasi, Ghana

Prosper Adiku
Institute for Environment and Sanitation Studies, College of Basic and Applied Sciences, University of Ghana, Accra, Ghana

Rahinatu Sidiki Alare
Faculty of Earth and Environmental Sciences, Department of Environmental Sciences, C.K. Tedam University of Technology and Applied Sciences, Navrongo, Ghana

Senyo Dotsey
Urban Studies and Regional Science, Gran Sasso Science Institute, L'Aquila, Italy

Abbebe Marra Wagino
Mendel University in Brno Project in Ethiopia, Addis Ababa, Ethiopia

Teshale W. Amanuel
Wondo Genet College of Forestry and Natural Resource, Hawassa University, Hawassa, Ethiopia

Eromose E. Ebhuoma
College of Agriculture and Environmental Sciences, Department of Environmental Sciences, University of South Africa (UNISA), Johannesburg, South Africa

Majoumo Christelle Malyse
Department of Geography, University of Dschang, Dschang, Cameroon

Olanike F. Deji
Obafemi Awolowo University, Ile Ife, Nigeria

S. A. Aderinoye-Abdulwahab and T. A. Abdulbaki
Department of Agricultural Extension and Rural Development, Faculty of Agriculture, University of Ilorin, Ilorin, Nigeria

Olumide A. Olaniyan and Ugbah Paul Akeh
National Weather Forecasting and Climate Research Centre, Nigerian Meteorological Agency, Abuja, Nigeria

Vincent O. Ajayi
West African Science Service Center on Climate Change and Adapted Land Use, Federal University of Technology, Akure, Ondo State, Nigeria

Department of Meteorology and Climate Science, Federal University of Technology, Akure, Nigeria

Kamoru A. Lawal
National Weather Forecasting and Climate Research Centre, Nigerian Meteorological Agency, Abuja, Nigeria
African Climate and Development Initiative, University of Cape Town, Cape Town, South Africa

Index

Printed in the USA
CPSIA information can be obtained
at www.ICGtesting.com
JSHW051351091023
49903JS00006B/106